14th Gravitational Waves Data Analysis Workshop 2010

(GWDAW-14)

Journal of Physics: Conference Series
Volume 243

Rome, Italy
26-29 January 2010

ISBN: 978-1-61738-961-0

Printed from e-media with permission by:

Curran Associates, Inc.
57 Morehouse Lane
Red Hook, NY 12571

CURRAN ASSOCIATES INC. proceedings.com

Some format issues inherent in the e-media version may also appear in this print version.

Printed by Curran Associates, Inc. (2010)

Institute of Physics
Dirac House, Temple Back
Bristol BS1 6BE UK

Phone: 44 1 17 929 7481
Fax: 44 1 17 920 0979

techtracking@iop.org

Additional copies of this publication are available from:

Curran Associates, Inc.
57 Morehouse Lane
Red Hook, NY 12571 USA
Phone: 845-758-0400
Fax: 845-758-2634
Email: curran@proceedings.com
Web: www.proceedings.com

TABLE OF CONTENTS

PREFACE

The 14th Gravitational Wave Data Analysis Workshop (GWDAW-14) is the last of a long series of annual meetings dedicated to the GW data analysis. This time the workshop was held at the Department of Physics of the University of Rome "Sapienza" and its scientic focus was on strengthening the connection among the gravitational wave and other astrophysical communities. Thus, a significant fraction of the workshop was dedicated to explore the potentialities of the multimessanger astronomy and in particular on the emerging neutrino observatories in conjunction with the GW observations. Moreover, several contributions were devoted to technical details of the analysis of real data from interferometric detectors, aimed at the improvement of the data quality for increasing the confidence in the detection of the first GW event. On the base of these techniques new GW upper limits on the strength of continuous signals from neutron stars and on stochastic background as the event rates of burst and inspiral signals have been set.

As chairman of this workshop, I would like to thank the members of the organizing and scientic committees and all the participants which have been the crucial actors of the workshop success. Some of the talks presented during the conference appear in the special issue of *Classical and Quantum Gravity*, while remaining talks from the symposium are published in this companion volume of *Journal of Physics: Conference Series*. The ensemble of all these contributions represents the most up-to-date papers on the topics covered by the meeting and, it provides valuable details about current work.

Finally , I would also like to thank the institutions and the sponsor that made this meeting possible:

- University of Rome *La Sapienza*
- Italian National Institute of Nuclear Physics – *INFN*
- Italian National Institute of Astrophyiscs – *INAF*
- University of Rome *Tor Vergata*
- University of *Sannio*
- *E4*–Computing Engineering s.p.a.

Fulvio Ricci
University of *La Sapienza* and *INFN*, Sezione di Roma, Rome, Italy
Chair, GWDAW-14 Scientic Organizing Committee

Open Questions in Astrophysically Triggered Gravitational Wave Searches

S. Márka[1] for the LIGO Scientific Collaboration and the Virgo Collaboration

[1]Columbia Astrophysics Laboratory, Columbia University in the City of New York, Pupin Physics Laboratories, New York, NY 10027 (USA)

Abstract. Sources of gravitational waves are often expected to also be observable through several other messengers, such as gamma rays, X-rays, optical, radio, and/or neutrino emission. Some of these channels are already being used in searches for gravitational waves with the LIGO-GEO600-Virgo interferometer network, and others are currently being incorporated into new searches. Astrophysical targets include gamma-ray bursts, soft-gamma repeaters, supernovae, and glitching pulsars. The simultaneous observation of electromagnetic or neutrino emission could be a crucial aspect for the first direct detection of gravitational waves. Information on the progenitor, such as trigger time, direction and expected frequency range, can enhance our ability to identify gravitational wave signatures with amplitudes close to the noise floor of the detector. Furthermore, combining gravitational waves with electromagnetic and neutrino observations will enable the extraction of scientific insight that was hidden from us before. The paper discusses the status of transient multimessenger detection efforts as well as intriguing questions that might be resolved in the future by advanced and third generation gravitational wave detectors.

1. Introduction

The astrophysics community expects direct detections from the advanced ground based gravitational wave detectors (aLIGO [1] and AdV [2, 3]) within the next decade [4, 5]. These detectors will achieve a tenfold increase in sensitivity around 100Hz over the initial detectors [4, 6]. Models and observations predict a detectable rate for binary neutron star coalescence between 0.4 to 400 events yearly [7].

The worldwide gravitational wave community is gearing up at the same time to evaluate the scientific reach and to produce a baseline design of the third generation of interferometric gravitational wave detectors (i.e.,Einstein Telescope (ET) [8]). These detectors will offer an order of magnitude better sensitivity around 100Hz than the advanced detectors and they could allow us to take advantage of routine detections of a wide range of gravitational wave emitters.

Gravitational wave transient events are believed to be often accompanied by other messengers [8–11], such as gamma-rays, optical photons, and/or neutrinos. It is expected that the information on the progenitor that we gain from observations through these other messengers is complementary to the information we may decipher from direct detection of gravitational waves [9]. Time, sky location, and information on the progenitor properties are often provided by multimessenger observations, that in turn could translate into more confident detection, efficient background rejection, and increased sensitivity. Some measurements are enabled only through multimessenger approaches [8, 12].

Frontier fields can be recognized easily; they often have more open questions than answers. This paper is organized as follows, to illustrate this point. In Section 2, we briefly review results of and motivation for recent multimessenger searches for transient gravitational waves. In Section 3, we provide a subset of the questions that may be answered by multimessenger searches with advanced and third generation ground based interferometric gravitational wave detector networks.

2. Brief Overview of the Status of Multimessenger Searches for Gravitational waves

The wordwide gravitatonal wave community has actively pursued multimessenger searches for more than a decade [9, 10, 13–18]. Most of observations are the results of externally triggered approaches [17, 19–26] that are summarized in this section.

Soft gamma repeaters (SGRs) [27] and anomalous X-ray pulsars (AXPs) are rare cosmic objects, less then two dozen are known in our Galaxy. Their putative engines are believed to be highly magnetized neutron stars, so called *magnetars* [28–30]. Their electromagnetic emission can correspond to single or rapidly repeated bursts of gamma-rays with isotropic energies between 10^{42} erg and 10^{46} erg. Some theories [31–36] predict that these bursts can also be accompanied by gravitational wave emission. Sporadic low energy flares are fairly common; they are observed hundreds of times per year [27].

On the other hand, so called "giant flares" with the highest energies are very rarely observed; we know only three instances [37–39]. Additionally, two of the giant flares were associated with quasiperiodic oscillations (QPOs) [40, 41], which might also be associated with gravitational wave emission. The search for gravitational waves from QPOs of the December 2004 giant flare of SGR 1806–20 [42] produced a null result but placed an observational limit on the associated gravitational wave energy. A separate search, targeting the instantaneous gravitational wave transient at the time of the giant flare [27], also returned a null result, and it also provided limits on the energetics. The analysis of a "storm" of 70 bursts of SGR 1900+14 [43] and an additional ~200 low-energy flares of SGR 1806–20 and SGR 1900+14 also extended our observational knowledge [27]. These searches triggered by Galactic SGR activity were probing neutron star f-modes experimentally. The isotropic gravitational wave energy limits reached as low as 2.4×10^{48} erg for some of the flares and emission models assuming ~10 kpc SGRs distance. A stacked search [44] for the gravitational wave signature of the SGR 1900+14 storm in 2006 provided an order of magnitude improvement in energy upper limits over the previous results [27].

Core collapse supaernovae are expected to be emitters of electromagnetic, neutrino, and gravitational wave signals [45]. While our knowledge of stellar core collapses is uncertain [46, 47], it is believed that these highly energetic events in the Galaxy can be associated with gravitational wave emission observable by advanced gravitational wave detector networks. Synergistic observations [48–54] of gravitational waves and low-energy neutrinos [55, 56] (up to a few 100 MeV) from core collapse supernovae might yield new information on the supernova engine and the dynamics of matter within. While most work on the phenomenology of the topic is yet to be done, some insight can already be gained from references [57–64] on the science that can be enabled by joint observations. A comprehensive review of the gravitational-wave signature of core-collapse supernovae is provided in [61] showing that a Galactic event is an interesting target for aLIGO, AdV and ET [8]. Present day neutrino detectors have a well established Galactic reach [49] for supernovae, that is a good match to the sensitivity of the advanced gravitational wave detectors [57, 59, 65–70]. While the expected rate of Galactic supernovae is very low (<1/20 years), future megaton class water-Cherenkov neutrino detectors [71–73] might allow a few Mpc reach ‡ that is a good match to the reach of third generation gravitational wave detectors [8].

High-energy neutrinos ($\gtrsim$ 100 GeV) can conceivably be emitted from shock-accelerated ultra-relativistic particles [74] e.g. in GRB outflows. Scientific possibilities related to joint observations with high energy neutrinos and gravitational waves are discussed in detail in [75]b.

Gamma-ray bursts (GRBs) [76] are believed to be associated with compact progenitors that are also possible emitters of gravitational waves [77]. Short GRBs (duration of burst is less than 2 s long) are believed to be associated with the merger [78]

‡ It is expected that the core-collapse supernovae rate within 5 Mpc is about 1/year.

of neutron star binaries, neutron star - black hole binaries, or rarely with giant flares [78] of extragalactic magnetars. While the gravitational wave signature of inspirals just preceding the merger is expected to be detectable to hundreds of Mpc with future detectors, the SGR reach probably remains confined to within our Galaxy. Long GRBs (>2s) are believed to be produced by energetic core-collapse supernovae [79–82]. While the model predictions for gravitational wave emission vary widely, it is likely that the advanced gravitational wave detectors can only have a reach that is well below 100 Mpc [56, 83–85]. Multiple searches for gravitational waves triggered by GRB observations were published by the LIGO and Virgo collaborations (e.g., recently [26, 86–88]). The search for the gravitational wave counterpart of GRB 070201 produced a non-detection and valuable scientific insight. The sky position error box of this short-hard GRB included the Andromeda galaxy (M31 at $\sim$770kpc). The result excluded binary compact coalescence in M31 at >99% confidence [86], a significant clue that the burst was due to a SGR giant flare in M31. The search for the gravitational wave counterpart of 22 short GRBs [87] produced lower limits, $\sim$O(several Mpc), on the binary progenitor distances. A model-independent search [88] covering 137 long and short GRBs yielded a null result, but with improved model-independent upper limits on the associated gravitational wave strain.

3. Some Open Questions that May be Answered by Multimessenger Observations with Gravitational Wave Detectors

While multimessenger astronomy with gravitational waves has already contributed to astrophysics, most of the exciting questions still remain to be answered (see e.g., [12, 89–91]). Indeed, it is a frontier field and it will take two generations of ground based gravitational wave detectors to answer a significant fraction of current open questions in the field. To clearly illustrate this point, we will list some of the questions in multimessenger astrophysics that might be resolved by advanced or third generation ground based gravitational wave detectors:

3.1. Some generic questions accessible through multimessenger observations

- What is the speed of gravitational waves, subluminal or superluminal [92, 93]?
- Does Einstein's theory of general relativity remain valid in the strong field regime [94]?
- Does gravity violate parity [95, 96]?
- Is there a new length scale beyond which general relativity is modified [97]?
- Which alternative gravity theories can be excluded experimentally [98]?
- How often can an unidentified electromagnetic transient be explained by a gravitational wave emitter?
- Is there a high redshift population of intermediate mass black holes [8]?

- Can gravitational waves help in explaining the origin of Ultra-Luminous X-ray binaries [8]?
- Can we search for new physics in the ultra-weak field regime [99]?
- Can a massive graviton serve as a cold-dark-matter candidate [92]?
- What fraction of the cosmic source's energy is emitted in the form of gravitational waves?
- Can gravitational wave detectors provide an early warning to electromagnetic observers to allow the detection of early light curves [9–12, 14, 89, 100]?
- Do gravitational measurements of distance agree with the concordance cosmology [101, 102]?
- What is the mass spectrum and spin distribution of black holes [103, 104]?
- Are there extra gravitational wave polarizations?
- ...

3.2. Some questions related to SGRs

- Is there a significant non-axisymmetric crust or core dynamics associated with SGR flares [34]?
- What is the precise origin of SGRs [31, 36]? (e.g., What is the mechanism for GW and EM emission and how are they correlated?)
- Is there a fundamental difference between giant and common SGR flares [33]?
- Do quark stars exist [36]?
- Can we exclude or confirm some of the SGR models [8, 31]?
- What is the origin of pulsar glitches?
- What is the composition and structure of neutron stars and their cores [31]?
- What is the tallest mountain that can be supported by neutron stars [36]?
- Can we use GW-EM observations to guide or EM+null GW results to distinguish the local extragalactic SGR contributions from the short GRB population?
- ...

3.3. Some questions related to core collapse supernovae

- What is the nature of gravitational collapse [61, 105, 106]?
- What is the relationship between the supernova progenitor and remnant (e.g., final mass and spin)?
- If the supernova remnant is not a black hole, how does it behave? (e.g., a transient hypermassive remnant with unstable modes or collapse to a BH?)
- What happens in a core collapse supernova before the light and neutrinos escape?

- What is the delay in between neutrinos and gravitational waves in a core collapse supernovae [107]?
- What is the role of anisotropic neutrino emission in supernovae [108]?
- What is the mass of a neutrino [109]?
- Can we see core collapse supernovae in gravitational waves that are not visible in neutrinos [8]?
- Is there an electromagnetically hidden population of core collapse events [110–112]?
- How many dynamical scenarios are associated with core collapse supernovae? Can we distinguish between them?
- Can pulsar birth kicks result in detectable gravitational waves?
- What is the time delay between the electromagnetic brightening and the core collapse of a supernova [113]?
- What are the properties of the core collapse supernova progenitor [47, 69, 114–119]?
- What is the role of the rotation and magnetic fields in stellar core collapse [106]?
- ...

3.4. Some questions related to GRBs

- What is the origin of long and short GRBs [8]?
- What is the precise dynamics of each GRB engine?
- Is there any longer-lasting central engine left over from the GRB explosion, and what's its nature [120–123]?
- Are there electromagnetically hidden populations of GRBs [8]?
- Does the hypothesized low luminosity GRB population exist [8]?
- Can we have direct inferences on the GRB jet parameters from gravitational waves?
- Can we estimate properties of the nuclear equation state using short GRBs [124]?
- Can we relate the luminosity distribution of GRBs to beaming and the central engine mechanism?
- Is it possible to construct a competitive Hubble diagram based on gravitational wave standard sirens [8, 101]?
- What is the relationship between the parameters of a compact binary system and it's electromagnetic and neutrino emission?
- What GRB progenitor models can we confirm or reject?
- Are there other (sub)classes of GRBs?
- Do choked GRBs exist [8]?
- What is the origin of choked GRBs?
- What is the cosmic population of choked GRBs?

- What are the engines producing high energy neutrino and gravitational wave emission together?
- What is the dynamics/energetics of joint high energy neutrino and gravitational wave emitters?
- What is the electromagnetic emission of binary neutron star coalescence [10]?
- What is the electromagnetic emission of a neutron star – black hole coalescence [10]?
- Is there any electromagnetic emission from binary black hole coalescence?
- What is the nature of XRFs and their relationship to long GRBs [8]?
- ...

It is expected that these lists of questions will evolve and grow substantially as we start answering the currently open questions. Naturally, one should not forget about the unexpected.

4. Acknowledgments

The authors gratefully acknowledge the support of the United States National Science Foundation for the construction and operation of the LIGO Laboratory, the Science and Technology Facilities Council of the United Kingdom, the Max-Planck-Society, and the State of Niedersachsen/Germany for support of the construction and operation of the GEO600 detector, and the Italian Istituto Nazionale di Fisica Nucleare and the French Centre National de la Recherche Scientifique for the construction and operation of the Virgo detector. The authors also gratefully acknowledge the support of the research by these agencies and by the Australian Research Council, the Council of Scientific and Industrial Research of India, the Istituto Nazionale di Fisica Nucleare of Italy, the Spanish Ministerio de Educación y Ciencia, the Conselleria d'Economia Hisenda i Innovació of the Govern de les Illes Balears, the Foundation for Fundamental Research on Matter supported by the Netherlands Organisation for Scientific Research, the Polish Ministry of Science and Higher Education, the FOCUS Programme of Foundation for Polish Science, the Royal Society, the Scottish Funding Council, the Scottish Universities Physics Alliance, The National Aeronautics and Space Administration, the Carnegie Trust, the Leverhulme Trust, the David and Lucile Packard Foundation, the Research Corporation, and the Alfred P. Sloan Foundation. The author is grateful to many colleagues for their comments and input on the text and content of the paper, especially to Emre Kahya, Alessandra Corsi, Richard O'Shaughnessy, Maxim Factourovich, Peter Shawhan, Zsuzsa Marka, Ray Frey, Andrea Viceré, Benoit Mours, Isabel Leonor, Peter Kalmus, Christian D. Ott, Eric Chassande-Mottin, Jonah Kanner, and Imre Bartos. This paper has been assigned LIGO Document Number LIGO-P1000068.

References

[1] LSC http://www.ligo.caltech.edu/advLIGO/

[2] VIRGO http://wwwcascina.virgo.infn.it/advirgo/

[3] Advanced Virgo Baseline Design document https://tds.ego-gw.it/ql/?c=6589 (*Preprint* VIR-0027A-09)

[4] The LIGO Scientific Collaboration 2009 *Reports on Progress in Physics* **72** 076901

[5] Whitcomb S E *et al.* 2009 *Astro2010 Decadal Survey White paper* https://dcc.ligo.org/cgi-bin/DocDB/ShowDocument?docid=1587

[6] VIRGO http://www.virgo.infn.it/

[7] LIGO Scientific Collaboration, Virgo Collaboration: J Abadie *et al.* 2010 *ArXiv e-prints* (*Preprint* 1003.2480)

[8] Chassande-Mottin E, Hendry M, Sutton P J and Márka S 2010 *ArXiv e-prints* (*Preprint* 1004.1964)

[9] Aso Y, Márka Z, Finley C, Dwyer J, Kotake K and Márka S 2008 *Class. Quantum Grav.* **25** 114039 (*Preprint* arXiv:0711.0107)

[10] Sylvestre J 2003 *Astrophys. Journal* **591** 1152–1156

[11] Stubbs C W 2008 *Class. Quant. Grav.* **25** 184033 arXiv:0712.2598

[12] Bloom J S *et al.* 2009 Astro2010 Decadal Survey Whitepaper: Coordinated Science in the Gravitational and Electromagnetic Skies arXiv: 0902.1527

[13] LIGO T and Collaborations V 2008 *Classical and Quantum Gravity* **25** 114051–+ (*Preprint* 0802.4320)

[14] Kanner J *et al.* 2008 *ArXiv e-print 0803.0312* **803** (*Preprint* 0803.0312)

[15] Clark J, Heng I S, Pitkin M and Woan G 2007 *Phys. Rev. D* **76** 043003–+ (*Preprint* arXiv:gr-qc/0703138)

[16] Hayama K, Mohanty S D, Rakhmanov M and Desai S 2007 *Classical and Quantum Gravity* **24** 681–+ (*Preprint* 0709.0940)

[17] Mohanty S D, Marka S, Rahkola R, Mukherjee S, Leonor I, Frey R, Cannizzo J and Camp J 2004 *Classical and Quantum Gravity* **21** 765–+

[18] van Putten M H, Levinson A, Lee H K, Regimbau T, Punturo M and Harry G M 2004 *Phys. Rev. D* **69** 044007–+ (*Preprint* arXiv:gr-qc/0308016)

[19] Collaboration T V 2007 *Classical and Quantum Gravity* **24** 671–+

[20] O'Shaughnessy R and LIGO Scientific Collaboration 2008 Just Add Photons: Astrophysics From Ligo External Triggers *AAS/High Energy Astrophysics Division* (*AAS/High Energy Astrophysics Division* vol 10) pp 06.04–+

[21] Kalmus P et al 2007 *Classical and Quantum Gravity* **24** 659

[22] The LIGO Scientific Collaboration 2005 *Phys. Rev. D.* **72** 042002 (*Preprint* gr-qc/0501068)

[23] Finn L S, Mohanty S D and Romano J D 1999 *Phys. Rev. D.* **60** 121101 (*Preprint* arXiv:gr-qc/9903101)

[24] The LIGO Scientific Collaboration 2008 *Phys. Rev. D* **77** 062004

[25] Dietz A 2008 GRB-triggered searches for gravitational waves in LIGO data *American Institute of Physics Conference Series* (*American Institute of Physics Conference Series* vol 1000) ed M Galassi, D Palmer, & E Fenimore pp 284–288

[26] The Virgo Collaboration 2008 *Classical and Quantum Gravity* **25** 225001

[27] Abbott B *et al.* 2008 *Phys. Rev. Lett.* **101** 211102 (pages 6) URL `http://link.aps.org/abstract/PRL/v101/e211102`

[28] Duncan R C and Thompson C 1992 *ApJ Lett.* **392** L9–L13

[29] Woods P M and Thompson C 2004 Soft gamma repeaters and anomalous x-ray pulsars: Magnetar candidates *Compact Stellar X-Ray Sources* ed Lewin W G H and van der Klis M (Cambridge: Cambridge Univ. Press) (*Preprint* `astro-ph/0406133`)

[30] Mereghetti S 2008 *A&A Rev.* **15** 225–287

[31] Corsi A and Owen B J 2009 A theoretical framework for LIGO-Virgo SGR searches: exploring Ioka's 2001 model Tech. Rep. LIGO DCC Document T0900242 and VIR-0028A-09 LSC and Virgo collaboration

[32] Ioka K 2001 *MNRAS* **327** 639–662 (*Preprint* `arXiv:astro-ph/0009327`)

[33] Horvath J E 2005 *Modern Physics Letters A* **20** 2799–2804 (*Preprint* `arXiv:astro-ph/0508223`)

[34] de Freitas Pacheco J A 1998 *A&A* **336** 397–401 (*Preprint* `arXiv:astro-ph/9805321`)

[35] Xu R X 2003 *Astrophys. J. Lett.* **596** L59–L62 (*Preprint* `astro-ph/0302165`)

[36] Owen B J 2005 *Phys. Rev. Lett.* **95** 211101 (*Preprint* `astro-ph/0503399`)

[37] Mazets E P *et al.* 1979 *Nature* **282** 587–589

[38] Hurley K *et al.* 1999 *Nature* **397** 41–43 (*Preprint* `astro-ph/9811443`)

[39] Hurley K, Boggs S E, Smith D M, Duncan R C, Lin R, Zoglauer A, Krucker S, Hurford G, Hudson H, Wigger C, Hajdas W, Thompson C, Mitrofanov I, Sanin A, Boynton W, Fellows C, von Kienlin A, Lichti G, Rau A and Cline T 2005 *Nature* **434** 1098–1103 (*Preprint* `arXiv:astro-ph/0502329`)

[40] Watts A L and Strohmayer T E 2006 *ApJ Lett.* **637** L117–L120 (*Preprint* `astro-ph/0512630`)

[41] Israel G L, Belloni T, Stella L, Rephaeli Y, Gruber D E, Casella P, Dall'Osso S, Rea N, Persic M and Rothschild R E 2005 *ApJ Lett.* **628** L53–L56 (*Preprint* `arXiv:astro-ph/0505255`)

[42] Abbott B *et al.* (LIGO Scientific Collaboration) 2007 *Phys. Rev. D* **76** 062003 (*Preprint* `astro-ph/0703419`)

[43] Israel G L, Romano P, Mangano V, Dall'Osso S, Chincarini G, Stella L, Campana S, Belloni T, Tagliaferri G, Blustin A J, Sakamoto T, Hurley K, Zane S, Moretti A, Palmer D, Guidorzi C, Burrows D N, Gehrels N and Krimm H A 2008 *ApJ* **685** 1114–1128 (*Preprint* `0805.3919`)

[44] The LIGO Scientific Collaboration 2009 *Astrophys. J.* **701** L68–L74 arXiv: 0905.0005

[45] Arnaud N, Barsuglia M, Bizouard M, Brisson V, Cavalier F, Davier M, Hello P, Kreckelbergh S and Porter E K 2004 *Astroparticle Physics* **21** 201–221 (*Preprint* arXiv:gr-qc/0307101)

[46] Janka H T, Langanke K, Marek A, Martínez-Pinedo G and Müller B 2007 *Phys. Rep.* **442** 38

[47] Ott C D 2009 *Class. Quant. Grav.* **26** 204015

[48] Scholberg K 2001 *Nuclear Physics B Proceedings Supplements* **91** 331–337 (*Preprint* arXiv:hep-ex/0008044)

[49] Fulgione W 2010 *J. Phys. Conf. Ser.* **203** 012077

[50] Fukuda S *et al.* 2003 *Nuclear Instruments and Methods in Physics Research Section A: Accelerators, Spectrometers, Detectors and Associated Equipment* **501** 418 – 462 ISSN 0168-9002 URL http://www.sciencedirect.com/science/article/B6TJM-4811KNK-B/2/efe9c5f7c6e3ae12e4764369e71910bf

[51] KamLAND URL http://www.awa.tohoku.ac.jp/KamLAND/

[52] for the LVD Collaboration F W 2006 *Memorie della Societa Astronomica Italiana Supplement* **9** "388–+"

[53] Alimonti et al G 2009 *Nuc. Inst. Meth. Phys. Res. A* **600** 568

[54] Ahrens J *et al.* 2004 *Astroparticle Physics* **20** 507 – 532 ISSN 0927-6505 URL http://www.sciencedirect.com/science/article/B6TJ1-49W31VF-1/2/7d423e933a84316c3be00b1c193cbd77

[55] Arnett W D, Bahcall J N, Kirshner R P and Woosley S E 1989 *Annu. Rev. of Astron. and Astrophys.* **27** 629–700

[56] Bethe H A 1990 *Rev. Mod. Phys.* **62** 801

[57] Marek A, Janka H T and Müller E 2009 *A&A* **496** 475

[58] Ott C D, Burrows A, Dessart L and Livne E 2008 *ApJ* **685** 1069

[59] Dimmelmeier H, Ott C D, Marek A and Janka H T 2008 *Phys. Rev. D* **78** 064056

[60] Röver C, Bizouard M, Christensen N, Dimmelmeier H, Heng I S and Meyer R 2009 *Phys. Rev. D* **80** 102004

[61] Ott C D 2009 *Class. Quantum Grav.* **26** 063001 arXiv:0809.0695

[62] Summerscales T Z, Burrows A, Finn L S and Ott C D 2008 *ApJ* **678** 1142

[63] Sagert I, Fischer T, Hempel M, Pagliara G, Schaffner-Bielich J, Mezzacappa A, Thielemann F and Liebendörfer M 2009 *Phys. Rev. Lett.* **102** 081101

[64] Abdikamalov E B, Dimmelmeier H, Rezzolla L and Miller J C 2009 *MNRAS* **392** 52

[65] Leonor I, Cadonati L, Coccia E, D'Antonio S, Di Credico A, Fafone V, Frey R, Fulgione W, Katsavounidis E, Ott C D, Pagliaroli G, Scholberg K, Thrane E and Vissani F 2010 *Class. Quant. Grav.* **27** 084019

[66] Kotake K, Iwakami W, Ohnishi N and Yamada S 2009 *ApJ Lett.* **697** L133

[67] Ott C D, Dimmelmeier H, Marek A, Janka H T, Hawke I, Zink B and Schnetter E 2007 *Phys. Rev. Lett.* **98** 261101

[68] Murphy J W, Ott C D and Burrows A 2009 *ApJ* **707** 1173

[69] Takiwaki T and Kotake K 2010 *Submitted to ApJ. ArXiv:1004.2896*

[70] Ott C D, Burrows A, Dessart L and Livne E 2006 *Phys. Rev. Lett.* **96** 201102

[71] Maricic J and the Lbne Dusel collaboration 2010 *Journal of Physics Conference Series* **203** 012109–+

[72] Nakamura K 2003 Hyper-kamiokande âĂŤ a next generation water cherenkov detector *Neutrinos and implications for Physics beyond the Standard Model* ed Shrock R (World Scientific)

[73] Spiering C 2008 Status and Perspectives of Astroparticle Physics in Europe arXiv: 0804.1500

[74] Waxman E and Bahcall J 1997 *Physical Review Letters* **78** 2292 URL `http://www.citebase.org/abstract?id=oai:arXiv.org:astro-ph/9701231`

[75] for the LIGO Scientific Collaboration and the Virgo Collaboration E C M 2010 *Journal of Physics Conference Series, Same Volume, submitted*

[76] Meszaros P 2006 *Reports of Progress in Physics* **69** 2259–2322 (*Preprint* `arXiv:astro-ph/0605208`)

[77] Kobayashi S and Mészáros P 2003 *Astrophys. Journal* **589** 861–870 (*Preprint* `arXiv:astro-ph/0210211`)

[78] Nakar E 2007 *Physics Reports* **442** 166–236 (*Preprint* `arXiv:astro-ph/0701748`)

[79] Galama T J *et al.* 1998 *Nature* **395** 670 arXiv:astro-ph/9806175

[80] Kulkarni S R *et al.* 1998 *Nature* **395** 663–669

[81] Campana S *et al.* 2006 *Nature* **442** 1008–1010 arXiv:astro-ph/0603279

[82] Cobb B E, Bailyn C D, van Dokkum P G and Natarajan P 2006 *Astrophys. Journal* **645** L113–L116 (*Preprint* `arXiv:astro-ph/0603832`)

[83] Woosley S E and Bloom J S 2006 *Ann. Rev. Astron. Astrophys.* **44** 507

[84] Davies M B, King A, Rosswog S and Wynn G 2002 *Astrophysical Journal* **579** L63–L66 (*Preprint* `arXiv:astro-ph/0204358`)

[85] Piro A L and Pfahl E 2007 *Astrophysical Journal* **658** 1173–1176

[86] The LIGO Scientific Collaboration 2008 *Astrophysical Journal* **681** 1419–1430

[87] and The Virgo Collaboration T L S C 2010 *The Astrophysical Journal* **715** 1453 URL `http://stacks.iop.org/0004-637X/715/i=2/a=1453`

[88] and The Virgo Collaboration T L S C 2010 *The Astrophysical Journal* **715** 1438 URL `http://stacks.iop.org/0004-637X/715/i=2/a=1438`

[89] Phinney E S 2009 Finding and Using Electromagnetic Counterparts of Gravitational Wave Sources arXiv:0903.0098

[90] Kulkarni S R and Kasliwal M M 2009 Astro2010 Decadal Survey Whitepaper: Transients in the local Universe arXiv:0903.0218

[91] Kocsis B, Frei Z, Haiman Z and Menou K 2006 *Astrophys. Journal* **637** 27–37 (*Preprint* arXiv:astro-ph/0505394)

[92] Dubovsky S L, Tinyakov P G and Tkachev I I 2005 *Phys. Rev. Lett.* **94** 181102

[93] Jacobson T and Mattingly D 2004 *Phys. Rev. D* **70** 024003

[94] Kocsis B, Haiman Z and Menou K 2008 *ApJ* **684** 870–887 (*Preprint* 0712.1144)

[95] Yunes N, O'Shaughnessy R, Owen B J and Alexander S 2010 *ArXiv e-prints* (*Preprint* 1005.3310)

[96] Alexander S and Yunes N 2009 *Phys. Rep.* **480** 1–55 (*Preprint* 0907.2562)

[97] Deffayet C and Menou K 2007 *The Astrophysical Journal Letters* **668** L143–L146 (*Preprint* 0709.0003)

[98] Desai S, Kahya E O and Woodard R P 2008 *Phys. Rev. D* **77** 124041–+ (*Preprint* 0804.3804)

[99] Kahya E O 2010 *ArXiv e-prints* (*Preprint* 1001.0725)

[100] Kulkarni S R and Kasliwal M M 2009 *Astro2010 Decadal Survey White paper* ArXiv:0903.0218

[101] Sathyaprakash B S, Schutz B and Van Den Broeck C 2009 Cosmography with the Einstein Telescope arXiv: 0906.4151

[102] Nissanke S, Hughes S A, Holz D E, Dalal N and Sievers J L 2009 *ArXiv e-prints* (*Preprint* 0904.1017)

[103] Fregeau J M, Larson S L, Miller M C, O'Shaughnessy R and Rasio F A 2006 *ApJ Lett.* **646** L135–L138 (*Preprint* arXiv:astro-ph/0605732)

[104] Brown D A, Brink J, Fang H, Gair J R, Li C, Lovelace G, Mandel I and Thorne K S 2007 *Physical Review Letters* **99** 201102–+ (*Preprint* arXiv:gr-qc/0612060)

[105] Kotake K, Sato K and Takahashi K 2006 *Reports on Progress in Physics* **69** 971–1143 (*Preprint* arXiv:astro-ph/0509456)

[106] Ott C D, Dimmelmeier H, Marek A, Janka H T, Zink B, Hawke I and Schnetter E 2006 *ArXiv Astrophysics e-prints*

[107] Pagliaroli G, Vissani F, Coccia E and Fulgione W 2009 *Physical Review Letters* **103** 031102–+ (*Preprint* 0903.1191)

[108] Buonanno A, Sigl G, Raffelt G G, Janka H and Müller E 2005 *Phys. Rev. D* **72** 084001–+ (*Preprint* arXiv:astro-ph/0412277)

[109] Arnaud N, Barsuglia M, Bizouard M A, Cavalier F, Davier M, Hello P and Pradier T 2002 *Phys. Rev. D* **65** 033010–+ (*Preprint* arXiv:hep-ph/0109027)

[110] Fischer T, Whitehouse S C, Mezzacappa A, Thielemann F K and Liebendörfer M 2009 *Astron. Astrophys.* **499** 1–15

[111] Sumiyoshi K, Ishizuka C, Ohnishi A, Yamada S and Suzuki H 2009 *ApJ Lett.* **690** L43–L46 (*Preprint* 0811.4237)

Journal of Physics: Conference Series **243** (2010) 012001 doi:10.1088/1742-6596/243/1/012001

[112] O'Connor E and Ott C D 2010 *Class. Quant. Grav. in press, arXiv:0912.2393 [astro-ph]*

[113] Matzner C D and McKee C F 1999 *ApJ* **510** 379 (*Preprint* arXiv:astro-ph/9807046)

[114] Gezari S, Dessart L, Basa S, Martin D C, Neill J D, Woosley S E, Hillier D J, Bazin G, Forster K, Friedman P G, Le Du J, Mazure A, Morrissey P, Neff S G, Schiminovich D and Wyder T K 2008 *ApJ Lett.* **683** L131

[115] Hamuy M 2003 *ApJ* **582** 905

[116] Smartt S J 2009 *ARA&A* **47** 63 (*Preprint* 0908.0700)

[117] Soderberg A M, Chakraborti S, Pignata G, Chevalier R A, Chandra P, Ray A, Wieringa M H, Copete A, Chaplin V, Connaughton V, Barthelmy S D, Bietenholz M F, Chugai N, Stritzinger M D, Hamuy M, Fransson C, Fox O, Levesque E M, Grindlay J E, Challis P, Foley R J, Kirshner R P, Milne P A and Torres M A P 2010 *Nature* **463** 513 (*Preprint* 0908.2817)

[118] Kasen D and Bildsten L 2009 *ArXiv:0911.0680*

[119] Fryer C L, Holz D E and Hughes S A 2004 *ApJ* **609** 288

[120] King A, O'Brien P T, Goad M R, Osborne J, Olsson E and Page K 2005 *The Astrophysical Journal Letters* **630** L113 URL http://stacks.iop.org/1538-4357/630/i=2/a=L113

[121] Perna R, Armitage P J and Zhang B 2006 *The Astrophysical Journal Letters* **636** L29 URL http://stacks.iop.org/1538-4357/636/i=1/a=L29

[122] Rosswog S 2007 *MNRAS* **376** L48–L51 (*Preprint* arXiv:astro-ph/0611440)

[123] Corsi A and Meszaros P 2009 *The Astrophysical Journal* **702** 1171 URL http://stacks.iop.org/0004-637X/702/i=2/a=1171

[124] Ferrari V, Gualtieri L and Pannarale F 2009 *ArXiv* (*Preprint* arXiv/0912.3692)

Joint searches for gravitational waves and high-energy neutrinos

Eric Chassande-Mottin for the LIGO Scientific Collaboration and the Virgo Collaboration

CNRS and Univ. Paris Denis Diderot, AstroParticule et Cosmologie (France)

E-mail: `ecm@apc.univ-paris7.fr`

Abstract. Many of the astrophysical sources and violent phenomena observed in our Universe are potential joint emitters of gravitational waves and high-energy cosmic radiation, in the form of photons, hadrons, and also neutrinos. This has triggered a collaborative analysis project between gravitational wave detectors and high-energy neutrino telescopes. In this article, we review some of the motivations for having pursuing science jointly and present the effort's status.

PACS numbers: 04.80.Nn,95.55.Vj,98.70Rz

Journal of Physics: Conference Series **243** (2010) 012002 doi:10.1088/1742-6596/243/1/012002

1. Introduction

This article reports the status of an on-going project [1, 2, 3] aiming at the joint detection of gravitational waves (GW) and high-energy neutrinos (HEN). GWs and neutrinos are similar in several ways. They are not absorbed‡ nor diffused by the interstellar medium or background radiation as opposed to high-energy photons. They are not deflected by extra-galactic magnetic fields as opposed to charged cosmic rays, so that the source location can be traced back from the direction of arrival. They interact weakly with the environment so that they can escape very dense media. Thanks to these properties, GW and HEN are both able to convey information from the core of the most violent astrophysical events in the Universe.

The basic motivation for the joint search of GW and HEN is the existence of potential common sources. In Sec. 2, we list those sources. We give a short description of the detectors involved in this project and examine the feasibility of the joint search in Sec. 3. In Sec. 4, we report on the development of the data analysis strategies.

2. Potential common cosmic sources of GW and HEN

Several conditions have to be met for an astrophysical object to yield significant emission of both GW and HEN. To generate GW, this object has to be compact (i.e, a massive and dense distribution of matter), and display a relativistic bulk motion. In order to be detectable by LIGO or Virgo, the periodic (i.e., orbital or spinning) or transient motion has to be characterized by a typical time-scale of order of tens of milliseconds, compatible with the detector sensitive frequency band. This must be accompanied by the ejection of relativistic baryons for the production of HENs. The astrophysical objects that satisfy those requirements include Galactic (e.g., connected to Soft Gamma Repeaters [4, 5] during their flaring episodes) and extra-Galactic associated to Gamma-Ray Bursts (GRBs). We will only discuss the latter in the following.

2.1. Gamma-ray bursts

Gamma-ray bursts (GRBs) are intense flashes of gamma-rays (see e.g., [6] for a recent review). They are associated with an exceptional energy release in the electromagnetic spectrum with an equivalent isotropic emission of order $E_{EM}^{iso} \sim 10^{50} - 10^{52}$ erg within a few seconds to few tens of seconds. GRBs are the most luminous events in the Universe observed up to today. They are observed to be isotropically distributed and usually located at cosmological distances.

The *fireball model* (see [7] for a review) provides a widely accepted phenomenological picture that accommodates the observations. A central engine ejects blobs of plasma at relativistic speeds. The inhomogeneities in the jet form shells that can propagate at different speeds. The shells collide eventually and electrons get accelerated during

‡ Strictly a very small fraction of the cosmic neutrinos is absorbed by the interstellar medium.

those shocks. They release their energy by producing synchrotron and inverse Compton γ§ which are finally observed as a GRB. If protons are also present in the jet, their interaction with the synchrotron photons can produce pions through the Δ-resonance whose decays produce (muon and electron) neutrinos. The production of such a burst of HENs would therefore be intimately related to that of the GRB [8].

There are two types of GRBs, phenomenologically distinguished by their duration. The short ($T \lesssim 2$ s) and long GRBs ($T \gtrsim 2$ s) are associated with different progenitors. The central engine of the former is thought to be a merger of neutron-star–neutron-star or black-hole–neutron-star binary systems [9], while there are indications that the latter is connected to the collapse of spinning massive star to a black-hole (so-called "collapsar" scenario [10]). Both are expected to be sources of GW bursts [6].

"Failed" GRBs — The fireball model has been successful in explaining a large fraction of the observations. There are however suspicions that the above mentioned process can sometimes follow a different path. To understand possible alternative scenarios, some insight into the physics of the fireball is needed. We need to explain why the GRB jets have to be ultra-relativistic, an issue that is usually referred to as the "compactness problem" [7].

Above a given critical energy, high-energy γ interact with photons of lower energies and form electron-positron pairs. The optical depth for this process can be shown to be $\propto f_p/R_e$ where f_p is the fraction of the photon pairs which satisfy the $e^- - e^+$ formation condition and R_e is the radius of the emission region. Ignoring relativistic effects, it is estimated to be of order 10^{13} for a typical GRB, by far too large for any photon to escape and hence the GRB to be observed. Relativistic effects blue-shift the photon energy (less photon pairs satisfy the pair formation condition, hence reducing f_p) and allow to stretch R_e (which is deduced from the observed time variability of the GRB) by a substantial factor. The net result is the reduction of the optical depth by a factor of$\Gamma^{4+2\alpha}$, with Γ the Lorentz factor, and $\alpha \approx 1$ to 2 the spectral index for high-energy photons [7]. The jet becomes optically thin for ultra-relativistic jets with$\Gamma \sim 100$ which is thus a requirement for the observation of GRBs.

Baryonic mass slows down the flow. Too much of it prevents the jet to be ultra-relativistic [7]. A slight baryonic pollution results in mildly relativistic jets with $\Gamma = O(1)$. In that case, the jet is optically thick and no GRB is observed. This scenario is usually referred to as "failed" GRB [11]. While the GRB does not occur in this case, GW and HEN emission remain.

The observation of such objects with conventional telescopes is challenging and GW and HEN provide an alternative. Although the rate of such events is unknown, it could be large [11] with a potential connection with the putative local under-luminous population of GRBs.

§ Here, γ refers to photons with energies in the gamma and hard X-ray parts of the electromagnetic spectrum.

3. Presentation of the detectors and feasibility of a joint GW and HEN search

The large-scale interferometric GW detectors Virgo and LIGO and the HEN telescopes ANTARES and IceCube are involved in a joint search project. For the sake of completeness, we present briefly the partner HEN telescopes. For more details, we refer the reader to the specific presentations included in these proceedings [12, 13].

Presentation of the HEN telescopes — ANTARES and IceCube are large-scale arrays of photo-multipliers connected to strings which instrument a large volume of sea water or ice. IceCube (located at the geographic South pole) is larger in size than ANTARES. The detector will reach the km^3 scale when completed with 80 strings. With 22 strings, ANTARES (located in the Mediterranean sea, near Marseilles) is a large-scale demonstrator for a future European km^3 detector in sea water.

Those detectors follow the same principle: a cosmic neutrino interacts with the detector environment and forms a relativistic muon. When the muon enters the water or ice, it generates a flash of Cerenkov light. This signature is detected by the optical modules if it occurs within the detector. The detection of the muon and the reconstruction of its track is based on local coincidences of the light hits compatible with the Cerenkov light front. Thanks to this detection principle, it is possible to detect neutrinos in the energy range above few tens of GeV. The typical reconstruction accuracy is about a degree with some difference whether ice or sea water is used.

The detectors are optimized to look downward. There is a large background coming from above associated with muons from cosmic-ray air showers and to a lesser extent from below due to the atmospheric neutrinos generated by the same process. The detection of cosmic neutrinos is thus possible only in the hemisphere opposite to upward-pointing vertical at the detector with an instantaneous sky coverage close to 2π sr. The location of ANTARES provides an annual sky coverage of about 3.5π sr, covering the Southern hemisphere and the Galactic center. The complementary IceCube detector is observing the Northern hemisphere, with full acceptance of 2π sr for the considered energy range (from a few tens of GeV to 100 TeV).

Because of the presence of background (due to atmospheric neutrinos and mis-reconstructed downward-going muons), the observation of an anisotropy in the directions of arrival or of a very energetic candidate event (with energy much larger than that of background) is required to differentiate it from background. A third possibility is the use of multi-messenger astronomy as discussed in Sec. 4.

Search feasibility — ANTARES and IceCube are operational and have been taking data during the last years in partial configurations. Several data sets overlap with the LIGO-Virgo science runs S5-VSR1 and S6-VSR2. It is likely that there will be more in the future as the program of upgrades appears nicely synchronized.

All the ingredients are joined together to make the search viable: there is a good sky coverage thanks to the complementarity of ANTARES and IceCube. Their sky coverage significantly overlaps with that of the GW detectors. The overlap is estimated to be roughly ~ 4 sr (~ 30 % of the sky) ‖ for each HEN telescope.

4. Development of the data analysis strategy

Investigations of the search feasibility have been performed [1, 2] with simulated data. The fundamental rationale for the coincidence search is that we are dealing with independent detectors, based on radically different physics but targeting the very same cosmic sources. The probability of an accidental time *and* spatial coincidence due to background can be set to very small value.

The results obtained for a hypothetical IceCube-LIGO network [1] and an IceCube-LIGO-Virgo network indicate that even if the individual observatories in the network provide several triggers a day, the false alarm rate for the combined detector network can be very low (e.g., $\sim 1/(500$ yr) for a network composed of the two LIGO detector and IceCube assuming a 1 second coincidence window).

The width of the time and spatial coincidence windows are particularly important ingredients. The uncertainties in the source model largely dominate the timing accuracy of each instrument. The width of the time coincidence window is thus determined by the maximum time delay between GW and HEN emission. In the case of GRBs, the production of HENs and γ are both expected to be comparable to the fireball lifetime. We can thus use gamma-ray observations to deduce the time delay [14]. As pictured in Fig. 1, the analysis of $\sim 10^3$ GRBs in the 4th BATSE catalog [15] shows that in 95% of the cases, the light curve has a duration T_{90} shorter than 150 seconds. Including the possibility of precursors (observed for a fraction of the GRBs), the worst cases where GW and HEN triggers are the most separated in time define the following time coincidence $[-350s, +200s]$.

The spatial coincidence window is essentially related to instrumental limitations. The error box for the reconstruction of the source sky position has been shown to be ~ 10 square degrees for GWs [16, 17]and $\sim$ square degrees for HEN [13, 12] with some dependency upon the characteristics of the event. Simulations are on-going to determine these more systematically and more precisely. We expect that the lowering of event selection threshold (allowed by the joint analysis) will worsen the average sky resolution.

The detailed architecture for the joint analysis pipeline is not completely defined. In addition to the approaches considered in the early works mentioned above [1, 2], we will explore the potential of the pipelines normally used for externally triggered searches (see [18] and references therein) such as the X-pipeline [19].

‖ The sky is considered "visible" to GW detectors where the combined antenna pattern is above the half-maximum.

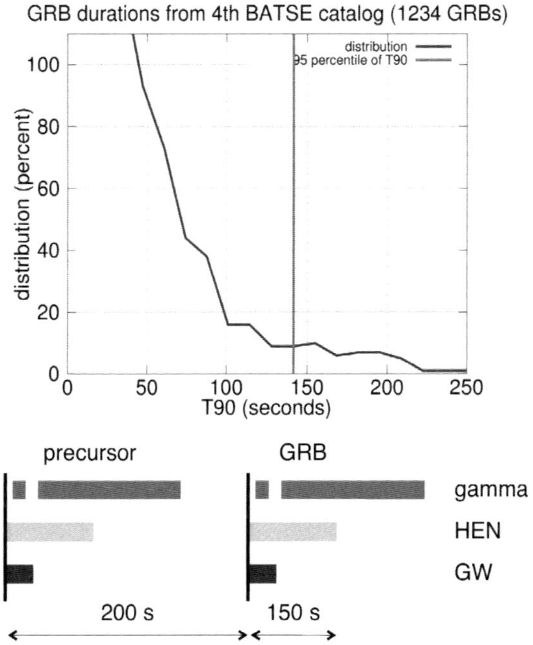

Figure 1. Time delay between GW and HEN — (top) This plot displays the distribution of the GRB duration taken from the 4th BATSE catalog. The duration of a GRB is thus determined to be shorter than 150 seconds with 95% confidence. (bottom) This figure shows the different components of the time delay between gravitational wave and high energy neutrino signals. Based on the available results, we cannot exclude the possibility of being able to detect GW, as well as HEN signals both from the precursor and the main GRB event. The resulting time coincidence window is $[-350s, +200s]$.

5. Concluding remarks

A collaboration between GW and HEN observatories has been started. A dedicated workshop held in Paris in May 2009 [20] provided a first occasion for the two scientific communities to meet and define a common plan. The working group joining GW and HEN contributors met face to face for the first time in Rome just before the GWDAW-14 workshop. Data exchange agreements have been signed by the involved parties allowing the more detailed definition of the joint data analysis procedure. In five to ten years from now, large improvements of the sensitivity and reach are likely both for the observations in GW and HEN channels. The project presented here is an important pioneering effort with existing data and a crucial path-finder effort for this "advanced" detector era.

Acknowledgments

The authors gratefully acknowledge the support of the United States National Science Foundation for the construction and operation of the LIGO Laboratory, the Science and Technology Facilities Council of the United Kingdom, the Max-Planck-Society, and the

State of Niedersachsen/Germany for support of the construction and operation of the GEO600 detector, and the Italian Istituto Nazionale di Fisica Nucleare and the French Centre National de la Recherche Scientifique for the construction and operation of the Virgo detector. The authors also gratefully acknowledge the support of the research by these agencies and by the Australian Research Council, the Council of Scientific and Industrial Research of India, the Istituto Nazionale di Fisica Nucleare of Italy, the Spanish Ministerio de Educación y Ciencia, the Conselleria d'Economia Hisenda i Innovació of the Govern de les Illes Balears, the Foundation for Fundamental Research on Matter supported by the Netherlands Organisation for Scientific Research, the Polish Ministry of Science and Higher Education, the FOCUS Programme of Foundation for Polish Science, the Royal Society, the Scottish Funding Council, the Scottish Universities Physics Alliance, The National Aeronautics and Space Administration, the Carnegie Trust, the Leverhulme Trust, the David and Lucile Packard Foundation, the Research Corporation, and the Alfred P. Sloan Foundation. This work was supported by the National Science Foundation under grant PHY-0847182, and by Columbia University in the City of New York. The author is grateful to many of their colleagues for their comments and input, especially to Imre Bartos for Fig. 1. This paper has been assigned LIGO Document Number LIGO-P1000067 and Virgo TDS number VIR-0356A-10.

References

[1] Y. Aso et al. Search method for coincident events from LIGO and IceCube detectors. *Class. Quantum Grav.*, 25:114039, 2008. arXiv:0711.0107.

[2] T. Pradier. Coincidences between gravitational wave interferometers and high energy neutrino telescopes. In *Proceedings of VLVnT08*, Toulon (France), 2008.

[3] V. Van Elewyck et al. Joint searches between gravitational-wave interferometers and high-energy neutrino telescopes: science reach and analysis strategies. *Int. J. Mod. Phys.*, D18:1655–1659, 2009. arXiv:0906.4957.

[4] K. Ioka. Magnetic deformation of magnetars for the giant flares of the soft gamma-ray repeaters. *MNRAS*, 327:639–662, 2001.

[5] K. Ioka, S. Razzaque, S. Kobayashi, and P. Mészáros. TeV-PeV neutrinos from giant flares of magnetars and the case of SGR 1806-20. *ApJ*, 633:1013–1017, 2005.

[6] G Vedrenne and J.-L. Atteia. *The brightest explosions in the Universe.* Springer, Berlin, 2009.

[7] T. Piran. Gamma-ray bursts and the fireball model. *Phys. Rep.*, 314(6):575–667, 1999.

[8] E. Waxman and J. Bahcall. High Energy Neutrinos from Cosmological Gamma-Ray Burst Fireballs. *Phys. Rev. Lett.*, 78:2292–2295, 1997.

[9] E. Nakar. Short-hard gamma-ray bursts. *Phys. Rep.*, 442:166–236, 2007.

[10] A. I. MacFadyen and S. E. Woosley. Collapsars: Gamma-Ray Bursts and Explosions in "Failed Supernovae". *ApJ*, 524:262–289, 1999.

[11] S. Ando and J. F. Beacom. Revealing the supernova gamma-ray burst connection with TeV neutrinos. *Phys. Rev. Lett.*, 95(6):061103–+, August 2005.

[12] F. Halzen. In *Proceedings of the 14th Gravitational Wave Data Analysis Workshop*, Rome, 2010.

[13] T. Pradier on behalf of the ANTARES Collaboration. The ANTARES Neutrino Telescope and Multi-Messenger Astronomy. In *Proceedings of the 14th Gravitational Wave Data Analysis Workshop*, Rome, 2010. arXiv:1004.5579.

[14] I. Bartos for the LIGO Scientific Collaboration and the Virgo Collaboration. Joint search between

a gravitational-wave detector network and high-energy neutrino detectors. AAS Meeting (Washington DC), 2010.

[15] W. S. Paciesas et al. The fourth BATSE gamma-ray burst catalog. *ApJS*, 122:465–495, 1999.

[16] S. Fairhurst. Triangulation of gravitational wave sources with a network of detectors. *New J. Phys.*, 11:123006, 2009.

[17] L. Wen and Y. Chen. Geometrical expression for the angular resolution of a network of gravitational-wave detectors. *Phys. Rev. D*, 81(8):082001, 2010.

[18] S. Màrka for the LIGO Scientific Collaboration and the Virgo Collaboration. Open questions in astrophysically triggered gravitational wave searches. In *Proceedings of the 14th Gravitational Wave Data Analysis Workshop*, Rome, 2010.

[19] Patrick J. Sutton et al. X-Pipeline: An analysis package for autonomous gravitational-wave burst searches. *New J. Phys.*, 12:053034, 2010.

[20] `http://www.gwhen-2009.org`.

Journal of Physics: Conference Series **243** (2010) 012003 doi:10.1088/1742-6596/243/1/012003

High frequency band sensitivity of the gravitational wave interferometer

N I Kolosnitzyn[1] and V N Rudenko[2]
[1]Shmidt Earth Physics Institute of RAS, Moscow, B. Gruzinskaya 10
nikkols_emto@mail.ru
[2]Sternberg Astronomical Institute, MSU, Moscow, Universitetskii prospect 13
rvn@sai.msu.ru

Abstract. It is shown that a reception frequency bandwidth of the large interferometer gravitational wave detectors in principle can be expanded in the high frequency side up to 100 kHz due to special "windows of reconstructed sensitivity" which have to appear in regions of EM-gravity resonances.
PACS: 04.80.Nn

1. Introduction

Recently the two long base laser interferometer gravitational wave detectors LIGO [1] and VIRGO [2] performed the first relatively short joint scientific run (S5 in LIGO, VSR-1 in VIRGO) with a total duration of 4.5 months [3]. After some new upgrade they have started (July 2009) with a second joint scientific run (S6, VSR-2).
General principles of such detectors were developed during of the last quarter of the past century and published in many papers, monographs and textbooks (see for example [4]).
A conventional theory of GW interferometers usually was considered with two main physical restrictions.
First,- it was accepted that a frequency of cosmic gravitational waves has to be much smaller the optical (a laser pump) frequency.
This restriction introduces a small parameter (λ_e / λ_g) << 1 that allows a using of a linear theory of gravitational wave interaction with the interferometer. Indeed for typical astrophysical GW sources this parameter does not exceed $\sim 10^{-10}$ and even for the high frequency tail of relic gravitational wave background it remains to be very small $\sim 10^{-4}$ or less [5].
Second,- it was considered the length of the interferometer base L was much less the gravitational wave length, or more exactly,- the interferometer light circulation frequency (or free spectral range frequency for FP cavity) $\nu_0 = c/2L$ was much larger the gravitational wave frequency i.e. $(\nu_g/\nu_0 << 1)$. For Michelson interferometers with FP cavities into its arms [1,2] the physical length has to be replaced by the effective value $L_{eff} = FL$ (where F is the *finesse* of FP cavity) with the limitation $L_{eff} \leq \lambda_g/2$. The last condition also means that such interferometer as a GW detector operates in a regime of "slow motions": the period of mirror's oscillation τ_g produced by the GW action has to be smaller the optical relaxation time of the FP cavity $\tau^* = F/\nu_0$.
Under this "slow motion" (or "large wave length") approximation one can define the interferometer antenna pattern (as a GW receiver) independently on a GW frequency.
At frequencies $\nu_g \geq \nu_0/F$ a detector transfer function begins to decrease $\sim \nu_0/F\nu_g$. It is just the reason of sensitivity degradation at the high frequency wing for the large GW interferometer detectors despite the optical shot noise spectrum density is kept at the constant level.
However it was noted in [6] that the interferometer transfer function at high frequencies might be reconstructed inside special narrow windows around frequencies of so called "gravitational - electromagnetic resonance" when GW frequency ν_g equals to an integer number of the interferometer free spectral range ν_0.
In this paper we continue this investigation more in detail. To simplify calculations we consider below the Michelson structure with multi-pass delay lines instead of FP cavities in the interferometer's arms. We

estimate the reconstructed sensitivity inside the resonance windows and also calculate a correspondent structure of the detector antenna pattern.

2. Optical delay line.

Electrical field of the laser beam going along the *x* direction through the optical delay line (ODL) in the presence of a gravitational wave can be read as [7, 8]

$$E^{(-)} = \frac{1}{2} r_1^{N-1} r_2^N \exp(i\omega_0 t)\left(\boldsymbol{h}\hat{D}^N \boldsymbol{A}\right) + c.c., \tag{1}$$

here $\omega_0 = 2\pi\nu_e$ $\omega_g = 2\pi\nu_g$ - optical and gravitational frequencies; r_1 r_2 - amplitude reflection factors of the front and far mirrors separated by the distance L; N – the number of round trips of the laser beam; the vectors $\boldsymbol{h}$ $\boldsymbol{A}$ are defined as:

$$\boldsymbol{h} = (1, \tfrac{1}{2} h_{11} \exp(i\omega_g t), \tfrac{1}{2} h_{11} \exp(-i\omega_g t)), \quad \boldsymbol{A} = (E_0, 0, 0). \tag{2}$$

If the gravitational wave falls down inclined in respect of the interferometer plane the matrix $\hat{D}$ looks like [7, 8]

$$\hat{D} = \exp(-i\mu)\begin{pmatrix} 1 & 0 & 0 \\ (k_0/2k_g)K & \exp(-i\varepsilon) & 0 \\ -(k_0/2k_g)K^* & 0 & \exp(i\varepsilon) \end{pmatrix}. \tag{3}$$

where the factor K is defined as

$$K(i\varepsilon) = \frac{1}{1+\alpha} + \frac{2\alpha}{1-\alpha^2} \cdot \exp(-i(1+\alpha)\varepsilon/2) - \frac{1}{1-\alpha} \cdot \exp(-i\varepsilon), \tag{4}$$

$$\mu = 2k_0 L, \quad \varepsilon = 2k_g L = \omega_g / \nu_0, \quad \nu_0 = c/2L, \quad \alpha = \sin\theta\cos\varphi. \tag{5}$$

here , are polar angles of the gravitational wave direction; the parameter α introduced above (5) is important for a delay line reaction.
A substitution (2), (3) into (1) results in

$$E^{(-)} = -r_1^{N-1} r_2^N E_0 e^{-iN\mu} \frac{1}{2} e^{i\omega_0 t} \left\{1 + h_{11} \frac{\mu}{4\varepsilon} \frac{\sin(N\varepsilon/2)}{\sin(\varepsilon/2)} \left[e^{i\omega t} e^{-i(N-1)\varepsilon/2} K_x - e^{-i\omega t} e^{i(N-1)\varepsilon/2} K_x^*\right]\right\} + c.c. \tag{6}$$

3. Michelson interferometer with delay lines in arms

The formula (6) presents the electrical field in the output light going along the *x* direction. At the output of a second delay line placed in *y* direction the electrical field will be described by a similar formula with following substitutions:

$$h_{11} \to h_{22,} \quad \mu_1 = k_0 2L_1 \to \mu_2 = k_0 2L_2, \quad \alpha = \sin\theta\cos\varphi \to \beta = \sin\theta\sin\varphi,$$
$$K_x \to K_y, \quad \varepsilon_1 = k_g 2L_1 \to \varepsilon_2 = k_g 2L_2, \quad \varepsilon_1 \cong \varepsilon_{2.} \tag{7}$$

The light beams from the both delay lines pass through a beam-splitter and interfere in two perpendicular outputs, so called the differential and common mode outputs. At an optical resonance a luminos-

Journal of Physics: Conference Series **243** (2010) 012003 doi:10.1088/1742-6596/243/1/012003

ity $I_\pm$ falling gown on registering photodiodes depends on the sum or difference of arm's electrical fields (its polarization is supposed to be in z direction)

$$I_\pm = \overline{(E_x \pm E_y)(E_x^* \pm E_y^*)} = \overline{E_x E_x^*} + \overline{E_y E_y^*} \pm \left(\overline{E_x E_y^*} + \overline{E_x^* E_y}\right) \qquad (8)$$

The upper bar reflects a time averaging performed by photodiodes filtering high frequency components. Low indexes in (8) mark ODL orientation; signs (±) correspond to the common (symmetrical) and differential (anti symmetrical) modes.

The substitution in (8) electrical fields of the arm's ODL (6), (7) results in the light intensity at output photo diodes

$$I_\pm = I_0\left(T_\pm(\Delta\mu) + T'_\pm h H_{ODL}(i\omega_g)\exp(i\omega_g t) + c.c.\right) \qquad (9)$$

$$T_\pm(\Delta\mu) = 1 \pm \cos(N\Delta\mu), \qquad T'_\pm \equiv \frac{dT_\pm}{d(\Delta\mu)} = \mp\sin(N\Delta\mu), \qquad (10)$$

$$H_{ODL}(\omega_g) = -i\frac{\mu}{4\varepsilon}\frac{\sin(N\varepsilon/2)}{\sin(\varepsilon/2)}\exp(-i(N-1)\varepsilon/2)[h_{11}K_x(\varepsilon) - h_{22}K_y(\varepsilon)].$$

(11)

here $K_{x(y)}$ is defined by (4); a transition from K_x to K_y corresponds to substitutions $\alpha \to \beta,\ L_1 \to L_2$. Other parameters are: $h = (h_+^2 + h_\times^2)^{1/2}$, $\Delta\mu = \mu_2 - \mu_1 = k_0 2(L_2 - L_1)$, $I_0 = r_1^{2N-2} r_2^{2N} I^*$, I^* - the input intensity of laser beam ; $T_\pm(\Delta\mu)$ characterizes a background luminosity at photo diodes. The second term in the right part of (9) describes variations of luminosity induced by gravitational wave. Its maximum can be achieved by the arm length tuning so that $L = 2(L_2 - L_1) = ($ e $/(4N))(1 + 2k)$, k is an integer number.

Formula (11) presents a transfer function of the interferometer as a GW detector. It depends on the frequency and angular variables included into the factor

$$F(\varepsilon,\theta,\varphi) = K_x(\varepsilon,\alpha)\frac{h_{11}}{h} - K_y(\varepsilon,\beta)\frac{h_{22}}{h}, \qquad (12)$$

This factor $F(\varepsilon,\theta,\varphi)$ can be considered as the detector antenna pattern. However in a general case this pattern will be frequency dependent.

For the particular case of orthogonal fall down (= =0, $h_{11} = -h_{22}$, $\varepsilon_1 \cong \varepsilon_2 = \varepsilon$, $K_x = K_y = 1 - e^{-i\varepsilon} = -i2e^{-i\varepsilon/2}\sin(\varepsilon/2)$) the frequency transfer function reduced to the well known form [7]

$$H_{ODL}(i\omega_g) = -i\frac{\mu}{2}\frac{\sin(N\varepsilon/2)}{\varepsilon/2}\frac{h_{11}}{h}\exp(-N\varepsilon/2). \qquad (13)$$

4. Long wave approximation

Conventional theory treats interferometer GW detectors as effective devices in the "long wave approximation" i.e. when the gravitational wave length is much larger the length of interferometer arms : $\lambda_g >> L$

; in this situation $\varepsilon \ll 1$. In the first order on ε the functions $K_x(\varepsilon,\alpha)$, $K_y(\varepsilon,\beta)$ do not depend on direction angles , and reduced to the value $K_x = K_y = i\varepsilon$.
Under transferring to the real form the expression (11) is factorized separating a frequency transfer function and angular antenna pattern.

$$H_{LW} = \frac{\mu}{4}\frac{\sin(N\varepsilon/2)}{\sin(\varepsilon/2)}F_0(\theta,\varphi), \qquad \varepsilon \ll 1. \tag{14}$$

$F_0(\theta,\varphi)$ is the angular antenna pattern:

$$F_0(\theta,\varphi) = (h_{11} - h_{22})/h = (h_+/h)(1+\cos^2\theta)\cos 2\varphi - (h_\times/h)2\cos\theta\sin 2\varphi. \tag{15}$$

Maximum of the transfer function (14) in the long wave limit $\varepsilon \to 0$ results in :

$$\max H_{LW} = \frac{\mu}{4}NF_0(\theta,\varphi) = \frac{\mu}{4}N\max\left(\frac{h_{11}-h_{22}}{h}\right). \tag{16}$$

5. Gravity-optical resonance (GOR)

Let's consider the transfer function (11) under an inclined gravitational wave falling down. We select frequencies ν_g equal an integer number of the circulation frequency $\nu_0 = c/2L$ so that $\nu_g = n\nu_0$. Under integer value of n (1, 2, 3...) the factor $\sin(N\varepsilon/2)/\sin(\varepsilon/2)$ in (11) has an uncertainty which can be easy solved. Taking into account that $\varepsilon/2 = \pi(\nu_g/\nu_0) = \pi n$, one comes to the following:

$$\frac{\sin(N\varepsilon/2)}{(\varepsilon/2)\sin(\varepsilon/2)} = \frac{\sin\pi Nn}{\pi n\sin\pi n} = \lim_{x\to 0}\frac{\sin\pi N(n+x)}{\pi n\sin\pi(n+x)} = \lim_{x\to 0}\frac{\cos\pi Nn}{\pi n\cos\pi n}\frac{\sin\pi Nx}{\sin\pi x} = (-1)^{-(N-1)n}\frac{N}{\pi n} \tag{17}$$

Then coming back to the formula (11) one gets in the maximum transfer function

$$H_{GOR}(n,\theta,\varphi) = -i\frac{\mu}{8\pi}\frac{N}{n}\left(K_x\frac{h_{11}}{h} - K_y\frac{h_{22}}{h}\right). \tag{18}$$

In the theory of spectral analysis the factor $\sin Nx/\sin x$, is known as the Bui-Ballot filter [9]: It filters with the same weights harmonics at frequencies equal integer number of π (i.e. $\nu = n\pi$).
But in our case according to formula (18) the weight of harmonic is inverse proportional of n .
In a close around of $\nu_g = n\nu_0$ ($n \ll N$) the transfer function symmetrically grows up to the value (18); one can call it as a "gravity-optical resonance" and expects an increasing sensitivity of GW-detection in these regions on a "white type" noise background.
Character of the transfer function (11) at resonance is defined mainly by the same Bui-Ballot factor with $x = (\nu_g/\nu_0)$ (other factors $F(\varepsilon,\theta,\varphi)$ and ε are changed slowly). Calculating the real part of (11) and omitting factors with module on the order of unit one can find the following estimate of the resonance transfer function

$$H_{GOR}(n,x,\theta,\varphi) = \frac{\mu}{8\pi}\frac{1}{n}\frac{\sin\pi N(x-x_0)}{\sin\pi(x-x_0)}F(n,\theta,\varphi), \quad |x| < x_0 = n \tag{19}$$

In the long wave approximation the transfer function (14) also contains the Bui-Ballot factor but in the low frequency region $\nu_g << \nu_0$ which one can call a "region of zero resonance".

It is worth to note a similar form of each resonance. Resonance number affects only the amplitude of transfer function (~ $1/n$).

6. Sensitivity enhancement in resonance zones

Appearance of gravity-optical resonances at frequencies $\nu_g = n\nu_0$ allows in principle to modify the sensitivity curve of large gravitational wave interferometers in its high frequency tail. For LIGO interferometers with the arm's length 4 km the circulation frequency of light is $\nu_0 = 37.5720$ *kHz*; the second and third zone corresponds to frequencies 75.040 kHz and 112.560 kHz. Widths of these zones are on the order of double zero zone width. It is interesting to estimate a factor of sensitivity increasing, using formulae for the frequency transfer function (14), (18). Conventional description of the GW-interferometer sensitivity is carried out in the form of the noise spectral density reduced to the instrument input. At the high frequency part the main noise source is the white photon shot noise depending only on the light intensity. Recalculation of this noise to the interferometer input is performed in fact through a simple division its spectral density on the instrument transfer function. Let's consider consequently the "long wave " and "gravity-optical resonance" regimes.

A) Long-wave regime.

In the frequency zone $\nu \le \nu_{\tilde{n}} = \tilde{n}/2NL$ signal intensity is estimated $\Delta I_s \approx I_0 N H_{LW} h$, where the frequency characteristic H_{LW} is defined by the formulae (14), (16).
In zone of the "effective reception" i.e. $\nu_g \le \nu_{\tilde{n}}$, $\varepsilon << 1$, one can get from (14)

$$|H_{LW}|^{-1} = |\frac{\mu}{4} \cdot N \frac{\sin x}{x}|^{-1} \approx \frac{1}{\pi}\frac{\nu_c}{\nu_e} \sim const\ , \quad x = N\varepsilon/2 = \pi\nu_g/\nu_c \le 1. \tag{20}$$

At more high frequencies $\nu_g > \nu_c$ the transfer function (14) is falling down. Its inverse value is growing up proportional ν_g:

$$|H_{LW}|^{-1} \approx \frac{\nu_g}{\nu_e} \tag{21}$$

At frequencies equaled integer numbers of circulation frequency this formula looks like

$$|H_{LW}|^{-1} \approx \frac{\nu_c}{\nu_e} n \cdot N \tag{22}$$

There is a low frequency plateau $\nu \le \nu_{\tilde{n}} = \tilde{n}/2NL$ with maximal sensitivity which has a linear decreasing for $\nu_g > \nu_c$.

B) Gravity-optical resonance regime.

At high frequencies $\nu_g > \nu_c$ under the condition $\nu_g = n\nu_0$ the transfer function (18) inverted to the interferometer input is estimated as

$$|H_{GOR}|^{-1} = |\frac{\mu}{8\pi}\frac{N}{n} F(n,\theta,\varphi)|^{-1} \approx 4\frac{\nu_c}{\nu_e} n \tag{23}$$

Ratio of the inverted transfer functions (22) to (23) presents a relative value of the noise spectral densities in regions of gravity-optical resonance and long wave regime

$$\frac{|H_{GOR}|^{-1}}{|H_{LW}|^{-1}} \approx \frac{4}{N} \tag{24}$$

Thus inside zones of gravity-optical resonance with n=1, n=2. a depressed sensitivity of LW regime is reconstructed partly reaching the level (23) i.e. the factor of increasing equals N/4.

7. Antenna pattern at gravity optical resonances.

Formulae of antenna patterns for the first gravity-optical resonances in a general case are too complex and we will consider below a particular case of direction cosines equality $\alpha = \beta = \pm(1/\sqrt{2}\cdot\sin\theta)$. Using the polarization formulae

$$h_{11} = h_+ \frac{1}{2}(\cos^2\theta - 1) - h_\times \cos\theta, \qquad h_{22} = h_+ \frac{1}{2}(\cos^2\theta - 1) + h_\times \cos\theta.; \tag{25}$$

supposing $h_+ = h_\times$; $\varepsilon/2 = \pi \cdot n$ and looking at (4) one has for $K_x(\varepsilon,\alpha)$

$$K_x(n,\alpha) = \frac{1}{1+\alpha} + \frac{2\alpha}{1-\alpha^2} e^{-i(1+\alpha)\pi\cdot n} - \frac{1}{1-\alpha} e^{-i2\pi\cdot n}.$$

Taking into account that $e^{-i2\pi\cdot n} = 1$, $e^{-i(1+\alpha)\pi\cdot n} = (-1)^n e^{-i\alpha\pi\cdot n}$ one can rewrite finally

$$K_x(n,\alpha) = -\frac{4\alpha}{1-\alpha^2} e^{-i\alpha\frac{\pi}{2}n} \begin{cases} \cos\left(\alpha\cdot\frac{\pi}{2}\cdot n\right) & n-\text{odd}, \\ i\sin\left(\alpha\cdot\frac{\pi}{2}\cdot n\right) & n-\text{even}. \end{cases}$$

A similar expression takes place for the second arm $K_y(n,\beta) = K_x(n,\alpha)$. At last coming from the complexity form to the real expression we find the antenna pattern as

$$F(n,\theta,\varphi) = \frac{4\alpha}{1-\alpha^2} \frac{h_{11} - h_{22}}{h} \begin{cases} \cos\left(\alpha\cdot\frac{\pi}{2}\cdot n\right) & n-\text{odd}, \\ \sin\left(\alpha\cdot\frac{\pi}{2}\cdot n\right) & n-\text{even}. \end{cases} \tag{26}$$

Here the antenna pattern specifics is defined only by the number of resonance n and does not depend on the number of round trip N in each arm.
Structures of antenna pattern (26) in the cross section $\varphi = \pi/4$, including the "zero resonance" (14) are given at figure 1a. The antenna pattern of "zero resonance" in the same cross section is reduced to $F_0(\theta,\varphi)|_{\alpha=\beta} = \sqrt{2}\cos\theta$. Its maximum corresponds to the gravitational wave with the normal falling down at the interferometer plane $\theta = 0$.

Journal of Physics: Conference Series **243** (2010) 012003 doi:10.1088/1742-6596/243/1/012003

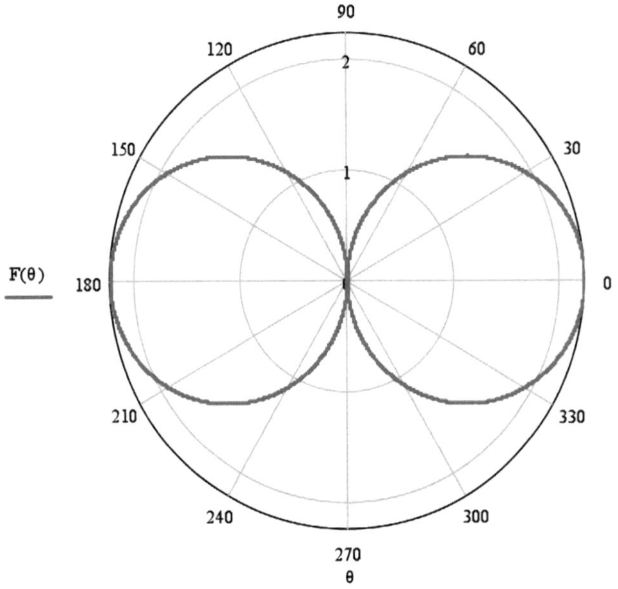

Figure 1a, n = 0, maxF = 2,221

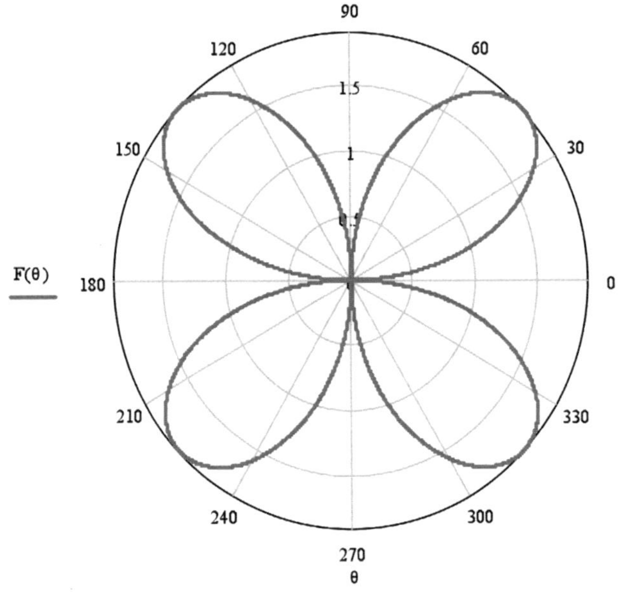

Figure 1b, n = 1, maxF = 1,889,

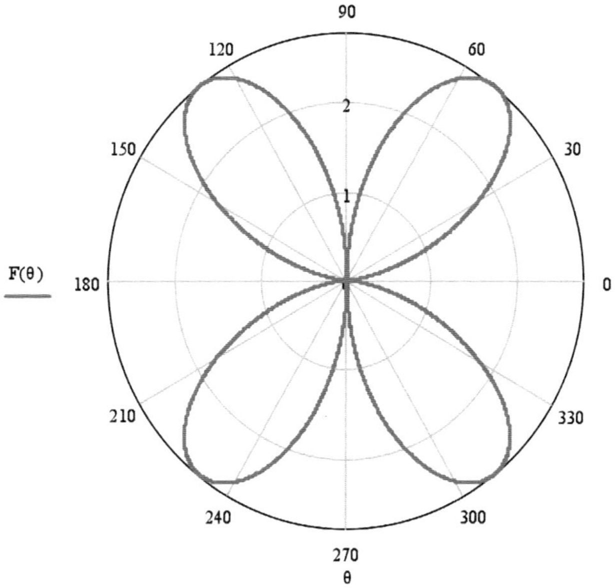

Figure 1c, n = 2, max(1/2)F = 1,385

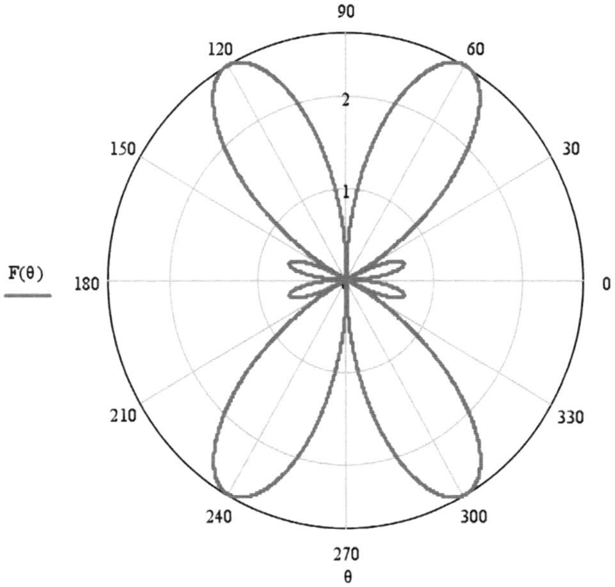

Figure 1d, n = 3, max(1/3)F = 0,984

8. Discussion

Antenna pattern configurations presented above demonstrate specific features of the gravity optical resonance – an angular shift of the maximum sensitivity depending on the number of resonance (or the receiver frequency) as well as a directivity (sharpness) enhancement with increasing a resonance number. It means that angular position of a source would be defined more accurate (compare with the antenna directivity in the low frequency limit) if it will be detected through the gravity-optical resonance with a large number n.
It is interesting to estimate the value of maximal "reconstructed sensitivity" inside the resonance zones. One can easily get such estimation for the case of dominant optical shot noise background (other noises are considered as much smaller). Then according to the simple "detection rule" $\Delta P_s \geq \Delta P_{fl}$ using (18) one estimates the minimal detectable metric variation as

$$P_0 H_{GOR} h \geq (2\hbar\omega P_0 \Delta f)^{1/2} \quad \text{or} \quad h_{\min} \geq \frac{\lambda \cdot n}{N}\sqrt{\frac{\hbar\omega \cdot \Delta f}{P_0}} \tag{27}$$

References

1. http://www.ligo.caltech.edu/
2. http://www.virgo.infn.it/
3. Abbott B P et al. 2009 In upper limit on stochastic gravitational-wave background of cosmological origin. *Nature* **460** 990-994
4. Saulson P R 1997 How an interferometer extracts and amplifies power from a gravitational wave. *Class. Quantum Grav.* **19** 2435-2454
5. Bisnovatyi-Kogan G S and Rudenko V N 2004 Very high frequency gravitational wave back ground in the Universe. *Class.Quantum Grav.* **21** 1-13
6. Rudenko V N and Sazhin M V 1980 Laser interferometer as a detector of gravitational waves *Kvantovaya electronica* **7** 2344-2355 (in Russian) translated in English 1980 as *"Sov.Quantum Electron."* **10** 1366-1373
7. Vinet J-Y 1986 Recycling interferometric antennas for periodic gravitational waves. *Physique.* **47** 639-643
8. Kolosnitin N I 1996 Maxwell equations and gravitational wave detection. *Gravitation & Cosmology.* **2** 262-266
9. Serebryanikov G and Pervozvanskii 1965 *Detection of covered periodicity* oscow

Tools for noise characterization in Virgo

T. Accadia[11], F. Acernese[5ac], F. Antonucci[8a], P. Astone[8a], G. Ballardin[2], F. Barone[5ac], M. Barsuglia[1], Th. S. Bauer[13a], M.G. Beker[13a], A. Belletoile[11], S. Birindelli[14a], M. Bitossi[7a], M. A. Bizouard[10a], M. Blom[13a], F. Bondu[14b], L. Bonelli[7ab], R. Bonnand[12], V. Boschi[7a], L. Bosi[6a], B. Bouhou[1], S. Braccini[7a], C. Bradaschia[7a], A. Brillet[14a], V. Brisson[10a], R. Budzyński[16b], T. Bulik[16cd], H. J. Bulten[13ab], D. Buskulic[11], C. Buy[1], G. Cagnoli[3a], E. Calloni[5ab], E. Campagna[3ab], B. Canuel[2], F. Carbognani[2], F. Cavalier[10a], R. Cavalieri[2], G. Cella[7a], E. Cesarini[3b], E. Chassande-Mottin[1], A. Chincarini[4], F. Cleva[14a], E. Coccia[9ab], C. N. Colacino[7ab], J. Colas[2], A. Colla[8ab], M. Colombini[8b], A. Corsi[8a], J.-P. Coulon[14a], E. Cuoco[2], S. D'Antonio[9a], V. Dattilo[2], M. Davier[10a], R. Day[2], R. De Rosa[5ab], G. Debreczeni[17], M. del Prete[7ac], L. Di Fiore[5a], A. Di Lieto[7ab], M. Di Paolo Emilio[9ac], A. Di Virgilio[7a], A. Dietz[11], M. Drago[15cd], V. Fafone[9ab], I. Ferrante[7ab], F. Fidecaro[7ab], I. Fiori[2], R. Flaminio[12], J.-D. Fournier[14a], J. Franc[12], S. Frasca[8ab], F. Frasconi[7a], A. Freise*, M. Galimberti[12], L. Gammaitoni[6ab], F. Garufi[5ab], M. E. Gáspár[17], G. Gemme[4], E. Genin[2], A. Gennai[7a], A. Giazotto[7a], R. Gouaty[11], M. Granata[1], C. Greverie[14a], G. M. Guidi[3ab], J.-F. Hayau[14b], H. Heitmann[14], P. Hello[10a], S. Hild**, D. Huet[2], P. Jaranowski[16e], I. Kowalska[16c], A. Królak[16af], N. Leroy[10a], N. Letendre[11], T. G. F. Li[13a], M. Lorenzini[3a], V. Loriette[10b], G. Losurdo[3a], E. Majorana[8a], I. Maksimovic[10b], N. Man[14a], M. Mantovani[7ac], F. Marchesoni[6a], F. Marion[11], J. Marque[2], F. Martelli[3ab], A. Masserot[11], C. Michel[12], L. Milano[5ab], Y. Minenkov[9a], M. Mohan[2], N. Morgado[12], A. Morgia[9ab], S. Mosca[5ab], V. Moscatelli[8a], B. Mours[11], I. Neri[6ab], F. Nocera[2], G. Pagliaroli[9ac], L. Palladino[9ac], C. Palomba[8a], F. Paoletti[7a,2], S. Pardi[5ab], M. Parisi[5b], A. Pasqualetti[2], R. Passaquieti[7ab], D. Passuello[7a], G. Persichetti[5ab], M. Pichot[14a], F. Piergiovanni[3ab], M. Pietka[16e], L. Pinard[12], R. Poggiani[7ab], M. Prato[4], G. A. Prodi[15ab], M. Punturo[6a], P. Puppo[8a], D. S. Rabeling[13ab], I. Rácz[17], P. Rapagnani[8ab], V. Re[15ab], T. Regimbau[14a], F. Ricci[8ab],

F. Robinet[10a], A. Rocchi[9a], L. Rolland[11], R. Romano[5ac], D. Rosińska[16g], P. Ruggi[2], B. Sassolas[12], D. Sentenac[2], L. Sperandio[9ab], R. Sturani[3ab], B. Swinkels[2], A. Toncelli[7ab], M. Tonelli[7ab], O. Torre[7ac], E. Tournefier[11], F. Travasso[6ab], G. Vajente[7ab], J. F. J. van den Brand[13ab], S. van der Putten[13a], M. Vasuth[17], M. Vavoulidis[10a], G. Vedovato[15c], D. Verkindt[11], F. Vetrano[3ab], A. Viceré[3ab], J.-Y. Vinet[14a], H. Vocca[6a], M. Was[10a], M. Yvert[11]

[1]AstroParticule et Cosmologie (APC), CNRS: UMR7164-IN2P3-Observatoire de Paris-Université Denis Diderot-Paris 7 - CEA : DSM/IRFU, France
[2]European Gravitational Observatory (EGO), I-56021 Cascina (PI), Italy
[3]INFN, Sezione di Firenze, I-50019 Sesto Fiorentino[a]; Università degli Studi di Urbino 'Carlo Bo', I-61029 Urbino[b], Italy
[4]INFN, Sezione di Genova; I-16146 Genova, Italy
[5]INFN, Sezione di Napoli [a]; Università di Napoli 'Federico II'[b] Complesso Universitario di Monte S.Angelo, I-80126 Napoli; Università di Salerno, Fisciano, I-84084 Salerno[c], Italy
[6]INFN, Sezione di Perugia[a]; Università di Perugia[b], I-06123 Perugia,Italy
[7]INFN, Sezione di Pisa[a]; Università di Pisa[b]; I-56127 Pisa; Università di Siena, I-53100 Siena[c], Italy
[8]INFN, Sezione di Roma[a]; Università 'La Sapienza'[b], I-00185 Roma, Italy
[9]INFN, Sezione di Roma Tor Vergata[a]; Università di Roma Tor Vergata, I-00133 Roma[b]; Università dell'Aquila, I-67100 L'Aquila[c], Italy
[10]LAL, Université Paris-Sud, IN2P3/CNRS, F-91898 Orsay[a]; ESPCI, CNRS, F-75005 Paris[b], France
[11]Laboratoire d'Annecy-le-Vieux de Physique des Particules (LAPP), IN2P3/CNRS, Université de Savoie, F-74941 Annecy-le-Vieux, France
[12]Laboratoire des Matériaux Avancés (LMA), IN2P3/CNRS, F-69622 Villeurbanne, Lyon, France
[13]Nikhef, National Institute for Subatomic Physics, P.O. Box 41882, 1009 DB Amsterdam[a]; VU University Amsterdam, De Boelelaan 1081, 1081 HV Amsterdam[b], The Netherlands
[14]Université Nice-Sophia-Antipolis, CNRS, Observatoire de la Côte d'Azur, F-06304 Nice[a]; Institut de Physique de Rennes, CNRS, Université de Rennes 1, 35042 Rennes[b], France
[15]INFN, Gruppo Collegato di Trento[a] and Università di Trento[b], I-38050 Povo, Trento, Italy; INFN, Sezione di Padova[c] and Università di Padova[d], I-35131 Padova, Italy
[16]IM-PAN 00-956 Warsaw[a]; Warsaw University 00-681 Warsaw[b]; Astronomical Observatory Warsaw University 00-478 Warsaw[c]; CAMK-PAN 00-716 Warsaw[d]; Białystok University 15-424 Bial ystok[e]; IPJ 05-400 Świerk-Otwock[f]; Institute of Astronomy 65-265 Zielona Góra[g], Poland
[17]RMKI, H-1121 Budapest, Konkoly Thege Miklós út 29-33, Hungary
*University of Birmingham, Birmingham, B15 2TT, United Kingdom
**University of Glasgow, Glasgow, G12 8QQ, United Kingdom

E-mail: elena.cuoco@ego-gw.it

Abstract. Several software tools were used to perform on-line and off-line noise analysis as a support to commissioning activities, to monitor the rate of glitches, the occurrence of non stationary noise, the presence of environmental contamination, the behavior of narrow spectral features and the coherence with auxiliary channels. We report about the use of these tools to study the main sources of identified noise: broadband, spectral lines and glitches. Plans for the upgrade of the tools will be presented, for example for lines identification purpose to let the scientists in control room do noise characterization in an easier way.

Journal of Physics: Conference Series **243** (2010) 012004 doi:10.1088/1742-6596/243/1/012004

1. Introduction

Several software tools were set up for the understanding and following up of the noise behavior for the Virgo gravitational-waves detector [?] and the results are reported on web pages. The goal is to have user-friendly tools to let the commissioning and data analysts easily characterize the data.

We want to have information on stationarity and gaussianity of the data, identify the spectral lines and characterize the transient noises. The idea was to have tools which, looking either at on-line or off-line data, produce results and plots which could be archived in files or MySQL database. We monitor

- the presence of transient noise in the data using different transient signal detection algorithms;
- the stationarity estimating the rms in frequency bands and spectrograms;
- the coherence of the gravitational wave channels and auxiliary channels to understand the source of spectral lines;
- the presence of continuous signals which appear as spectral peaks or lines in the frequency domain, trying to follow the temporal behavior of wandering lines.

2. Transient noise signals

Signals which last for a very short period of time are referred to as transient signals. The monitoring of transient noise signals is fundamental for the characterization of the data, since the gravitational waves signal produced by a supernovae explosion or a coalescing binaries event is a transient signal and we need to disentangle noise events by genuine Gravitational Waves (GW) signals. We need to understand how the transient noise can spoil the data and the path it follows to enter in the GW channel. For this reason we monitored these events in the GW channel and in a large set of auxiliary and environmental channels.

Event Trigger Generator (ETG) tools were used to produce lists of triggers for noise events. Two of them, WDF (Wavelet Detection Filter) [?] and VirgoHACR [?], produce triggers which are sent to a MySQL database, while Omega-pipeline [?] and OutlierMoni [?] write triggers in ASCII files on disk. We report in figure ?? an example of the plots which could be seen on the web. Moreover we produced summary pages for the last 1 hour, 8 hours and 24 hours, where we report frequency-time plots for the triggers and glitch rate values. The most energetic events are followed up in the GW channel and auxiliary channel in order to understand the origin of the event. The Omega-Scan [?] tool scans in the time-frequency domain a set of auxiliary and environmental channels and a report is automatically produced on the web.

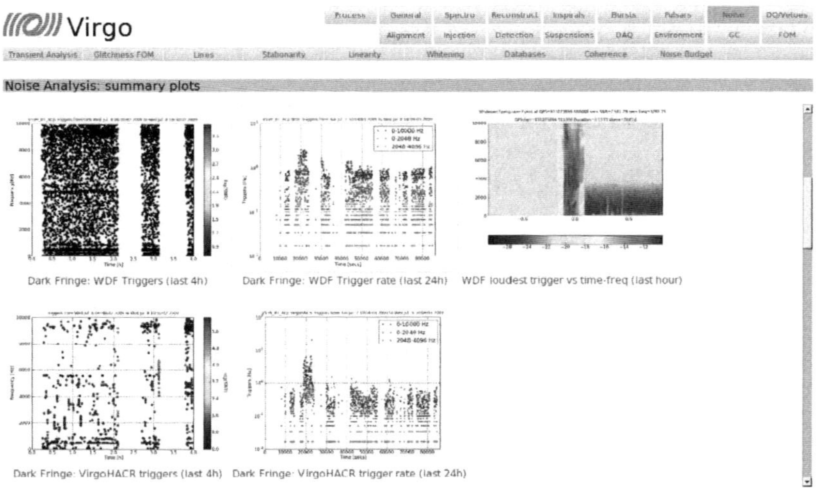

Figure 1. Example of monitor web pages: transient events detection, glitch rates and follow up of most energetic events

3. Slow non stationary noise

A tool was developed to run on-line in the Virgo environment (NonStatMoni) [?] to monitor how the noise is modulated in time. It can compute the root mean square (RMS) in several frequency bands, with different spectral resolutions for different channels.

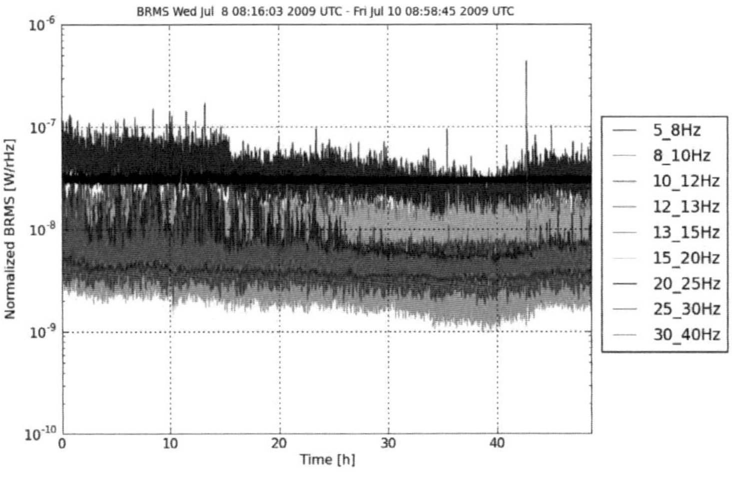

Figure 2. Example of monitors: Band-limited RMS in different frequency bands. Here is plotted the BMRS for uncalibrated GW channel. The low frequency part of the detector sensitivity band is limited by seismic noise. In this plot it is evident that in the first 10 hours there is higher noise in the low frequency band. This is due to excess of seismic noise.

Figure ?? shows an example of the plots produced for the low frequency band [?].

In the plot reported, there was an excess of seismic noise in the first 10 hours; this was evident as higher values of Bandlimided Root Mean Square (BMRS) in the low frequency bands. The excess of seismic noise causes up-conversion via scattering of diffused light on the optical benches [?].

4. Lines identification

We refer to continuous wave signal as the signals which have constant frequency and infinite time duration. To identify continuous wave signals in the GW channel, an algorithm has been developed to separate the contribution due to the broad band noise from the one due to narrow lines (LineMonitor) [?]. The algorithm works in two main steps: first it estimates the noise floor and afterward it searches for lines and estimates their parameters. Therefore it allows one to track separately the evolution in time of the power in lines and in the bulk of the noise spectrum. It is used in the on-line LineMonitor process for high resolution search for lines in the dark fringe spectrum.

In addition, we are developing a framework for the identification of the lines which makes use of some of the tools developed and used by the Continuous Waves (CW) search group [?]. The CW code (written in C) has been adapted in order to run as an "event finder" in either the calibrated GW channel data or uncalibrated GW channel and the set of auxiliary channels. The analysis procedure goes as follows:

- program looks for transient events in the time domain, which would affect the subsequent analysis in the frequency domain.
- time domain events are removed from the data, thus obtaining a "cleaned" data set (for the details see ref. [?]);
- program estimates the average spectrum, by means of an auto-regressive (AR) technique [?];
- From the average AR spectrum the events in the frequency domain higher than a given threshold are identified;
- Besides, the removed time and frequency domain events, together with their main parameters are archived in an ASCII file and could be analyzed for transient signal detection goals.

Some preliminary scripts have been run to read and analyze the daily data, produce time-frequency and persistency plots and publish them on a web page [?] or to load results in a MySQL database. Once the data are in the MySQL database, they can be retrieved by means of the APIs available for various platforms and operative systems.

Figure **??** presents two example plots obtained with the data analyzed by the event finder and retrieved from the database. The plots show the time-frequency diagrams of the calibrated GW channel data (left) and of an auxiliary channel (Em_SETODE01, right, which is a seismic sensor on the vacuum chamber containing the bench where the photo-diodes for the GW signal detections are).

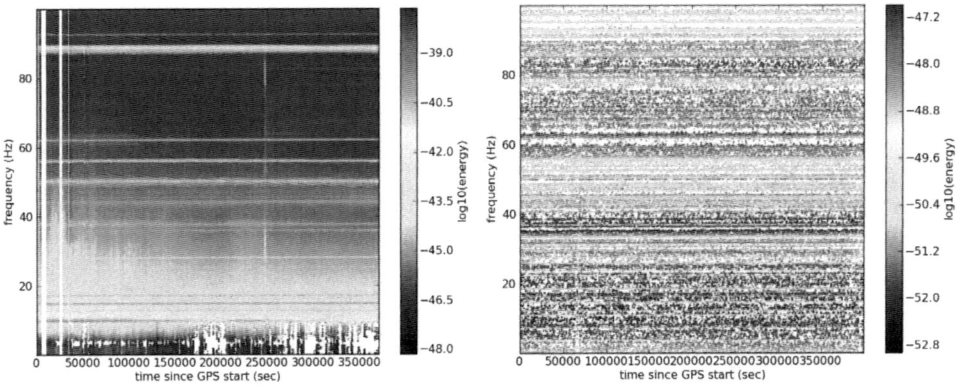

Figure 3. Time-frequency diagrams of the calibrated GW channel data (left) and of one seismic auxiliary channel (right). The color scale is based on the logarithm of energy of the frequency event.

5. Coherence

To understand the source of lines which enter in the GW channel, we automatically compute coherence between the dark fringe and all existing channels (about 1000), averaged over 15 minutes of data in about 1 h.

It run periodically (three times per hour) to always give updated results, which are summarized in a web page [?]. It produces links to coherence plots of all analyzed channels (see figure ??) and a list of the 10 most coherent channels for each frequency bin (see figure ??). We plan to archive such results in a MySQL database in order to

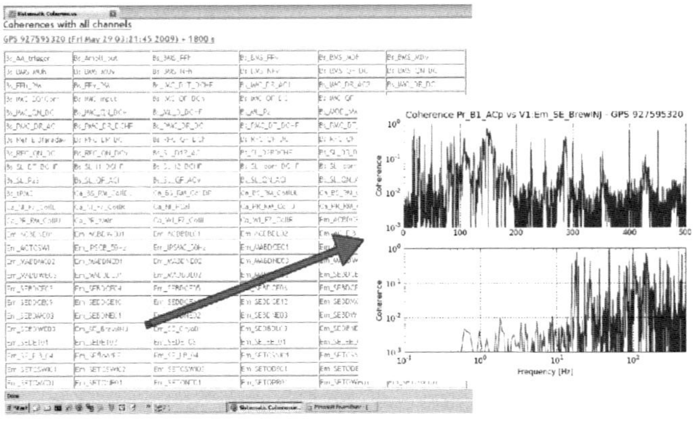

Figure 4. Table of coherence between GW channel and auxiliary channels

retrieve the information on user demand using simple web interfaces.

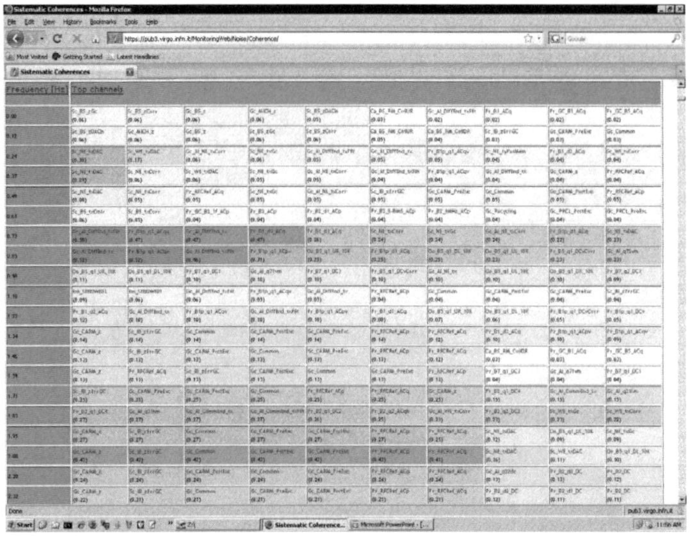

Figure 5. The most coherent channel organized by frequency bin.

6. Conclusions

We set up monitoring tools to characterize data and produce in-time results useful for data analysis and commissioning teams. The framework is still not complete. For example, we need to set up tools for the understanding and detection of non linear noise. We worked in the last months to set up of a new line monitoring tools which could work also on auxiliary channels, with the aim to have automatically the lines which are common to the calibrated GW channel and auxiliary ones. The ultimate goal is to follow the wandering lines using cross correlation between the GW channel and noise monitor channels.

Moreover we are working on a web interface for the noise data base which could let the users query easily the archived information, produce plots or launch scripts to follow up of the events or coherence.

References

[1] http://www.virgo.infn.it.
[2] E. Cuoco, Tech. Rep. VIR-NOT-EGO-1390-308 (2005).
[3] G. Vajente, Tech. Rep. VIR-0002A-08(2008).
[4] G. M. E. K. S Chatterji, L Blackburn, *Class.Quant.Grav* **21**, S1809 (2004).
[5] E. Chassinde-Mottin, Tech. Rep. VIR-0284A-05.
[6] G.Vajente, *NonStatMoni technical description*, Tech. Rep. VIR-004A-08.
[7] https://pub3.virgo.infn.it/MonitoringWeb/Noise/NonStatMoni.
[8] T. Accadia et al *Noise from scattered light in Virgo's second science run data* this proceeding.
[9] G. Vajente, *LineMonitor*, Tech. Rep. VIR-0564A-09V.
[10] F. Acernese *et al* (Virgo Coll.), *Cleaning the Virgo sampled data for the search of periodic sources of gravitational waves*, Class. Quantum Grav. 26 (2009).
[11] S. M. Kay *Modern Spectral Estimation* Prentice-Hall, Englewood Cliff, New Jersey 1995

[12] http://wwwcascina.virgo.infn.it/MonitoringWeb/Pulsars/index.html.
[13] https://pub3.virgo.infn.it/MonitoringWeb/Noise/Coherence.

Journal of Physics: Conference Series **243** (2010) 012005 doi:10.1088/1742-6596/243/1/012005

Used Percentage Veto for LIGO and Virgo Binary Inspiral Searches

Tomoki Isogai (for the LIGO Scientific Collaboration and the Virgo Collaboration)

Physics and Astronomy, Carleton College, Northfield, MN 55057, USA

E-mail: `isogait@carleton.edu`

Abstract. A challenge for ground-based gravitational wave detectors such as LIGO and Virgo is to understand the origin of non-astrophysical transients that contribute to the background noise, obscuring real astrophysically produced signals. To help this effort, there are a number of environmental and instrumental sensors around the site, recording data in "channels". We developed a method called the used percentage veto to eliminate corrupted data based on the statistical correlation between transients in the gravitational wave channel and in the auxiliary channels. The results are used to improve inspiral binary searches on LIGO and Virgo data. We also developed a way to apply this method to help find the physical origin of such transients for detector characterization. After identifying statistically correlated channels, a follow-up code clusters coincident events between the gravitational wave channel and auxiliary channels, and thereby classifies noise by correlated channels. For each selected event, the code also gathers and creates information that is helpful for further investigations. The method is contributing to identifying problems and improving data quality for the LIGO S6 and Virgo VSR2 science runs.

Journal of Physics: Conference Series **243** (2010) 012005 doi:10.1088/1742-6596/243/1/012005

1. Introduction

LIGO [1] and Virgo [2] started science runs called S6 and VSR2, respectively, in July 2009 (GEO 600 [3] is also participating in this run). The LIGO Scientific Collaboration (LSC) and Virgo Collaboration are analyzing the data for direct observations of gravitational waves (GWs) of cosmic origin. The LIGO and Virgo detectors are, however, susceptible to various kinds of noise. Their sensitivity to GWs depends on how well such noise can be controlled. Naturally much effort in LIGO and Virgo goes into understanding noise and characterizing the data obtained. To maintain the quality of data and minimize the effect of noise transients, or so called "glitches", we exclude, or more commonly referred to as "veto", contaminated data that are not suited to be analyzed for GW detection. Since a number of environmental and instrumental noise sources can potentially couple into the main detector output, many of them are continuously recorded along with the data from the main detector output. The measurement points for time-series data from various monitors are referred to as "channels". The channels include interferometer control signals such as length and angle control signals for optical cavities and physical environmental monitors (PEMs) such as seismometer, magnetometer and microphones around the interferometer.

Conventionally, we have only vetoed noise whose cause is fairly well understood with the consequence that many glitches remained in the science data due to their unknown origin [4, 5]. To eliminate the remaining glitches, we have developed a method called the used percentage veto (UPV). It utilizes the information from various instrumental and PEM channels, and uses their statistical correlation with the GW channel to exclude noisy times from our science data, as opposed to using knowledge about the deterministic causes of the noise. The result of this method is being used to improve the search for GWs from the inspiral of compact binary systems. In addition, the statistical data from UPV is used to help identify the physical cause of noise.

In section 2, we describe the method of UPV. Section 3 explains the application of UPV for detector characterization. Finally, a summary is given in section 4.

2. Used Percentage Veto

UPV finds auxiliary channels that are statistically correlated with the GW channel. To measure the correlation, we use KleineWelle (KW) triggers [6] (UPV can be applied to any trigger generation algorithm; we use KW for its availability on many channels and its relatively low latency). KW is a multiresolution method to find and characterize transients in an input timeseries. KW makes use of the dyadic wavelet transform to search for regions of excess energy in the time-scale decomposition. KW is applied to instrumental and environmental channels to find brief periods of excess power in each channel in near real time. Each trigger has a peak time and a significance (which describes the amplitude of triggers). The UPV code finds time-coincident triggers between the GW channel and an auxiliary channel with a ± 1s coincidence window using triggers that 1) have at least a KW significance of 50 and 2) are in time segments after severely corrupted times with well-known causes are excluded [4, 5, 7]. Starting from a KW significance of 50, the code raises the threshold at a step interval of 50 up to 5000 (these values are chosen empirically considering accidentals and computational cost), and at each threshold the code calculates various metrics, or "figures of merit", that measure how well the channel would work as a veto. These figures of merit are defined as follows:

- Used Percentage

$$\text{Used Percentage}(\rho) \equiv \frac{100 \times N^{\text{aux}}_{\text{coinc}}(\rho)}{N^{\text{aux}}_{\text{Total}}(\rho)} \qquad (1)$$

where $N^{\text{aux}}_{\text{coinc}}$ is the number of the auxiliary channel KW triggers coincident with the GW channel above the KW significance threshold ρ and $N^{\text{aux}}_{\text{Total}}$ is the number of the total auxiliary channel KW triggers above the KW significance threshold ρ. This value tells how well the channel is correlated with the GW channel. For an ideal veto, the used percentage would be 100%, but in reality the used percentage remains under 50% for most vetoes. UPV requires a used percentage of at least 50% with $N^{\text{aux}}_{\text{coinc}}$ larger than 10 over the period of time analyzed (a week for S6/VSR2) for channels to be a veto candidate (see Fig. 1). These parameters are determined empirically to ensure sufficient correlation between the GW and auxiliary channel.

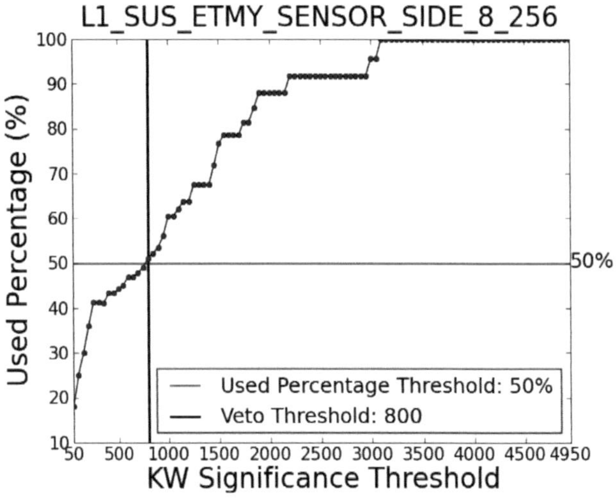

Figure 1: A plot showing used percentage as a function of KW significance for a particular suspension signal. UPV picks a significance threshold where the used percentage exceeds 50%, shown as a horizontal line in the plot.

- Efficiency

$$\text{Efficiency}(\rho) \equiv \frac{100 \times N^{\text{GW}}_{\text{vetoed}}(\rho)}{N^{\text{GW}}_{\text{Total}}} \qquad (2)$$

where $N^{GW}_{vetoed}(\rho)$ is the number of vetoed GW channel triggers at the KW threshold ρ and N^{GW}_{Total} is the total number of the GW channel triggers analyzed. This value tells how well the veto would eliminate glitches in GW channel.

- Dead Time

$$\text{Dead Time}(\rho) \equiv \frac{100 \times T_{vetoed}(\rho)}{T_{Total}} \tag{3}$$

where T_{vetoed} is the time vetoed at the KW threshold ρ and T_{Total} is the total time analyzed. A good veto has a high efficiency with a low dead time. Thus, the ratio

$$\frac{\text{Efficiency}(\rho)}{\text{Dead Time}(\rho)} \tag{4}$$

is often used to characterize the veto (see Fig. 2).

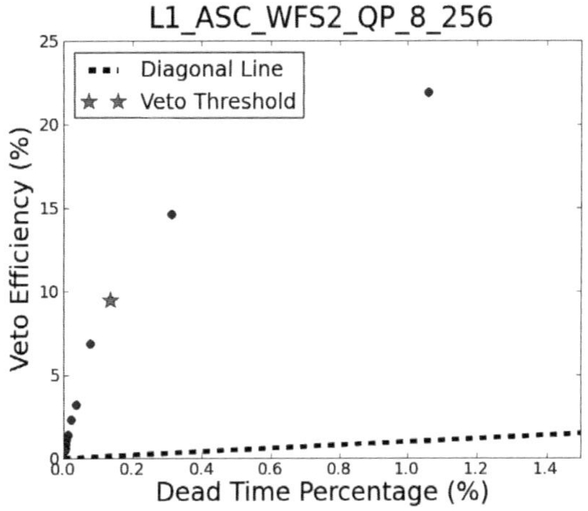

Figure 2: A plot showing the veto efficiency vs dead time for a particular wave front sensor. The dots correspond to each threshold, and the dashed line is the diagonal (efficiency = deadtime). A "good veto" has a high efficiency and a low dead time, so we would expect the points to be well above the diagonal line. In this case, we can see that the point at our threshold of choice (star) makes a very effective veto.

- Random Used Percentage

$$\text{Random Used Percentage}(\rho) \equiv \frac{\text{Dead time}(\rho) \times N^{GW}_{total}}{N^{aux}_{total}(\rho)} \tag{5}$$

This is the used percentage we expect when triggers in auxiliary channel are randomly distributed, i.e. not correlated with triggers in the GW channel. Therefore, the ratio

$$\frac{\text{Used Percentage}(\rho)}{\text{Random Used Percentage}(\rho)} \tag{6}$$

tells how well the channel is correlated with the GW channel above the random chance value. Note that this value is related to Efficiency/Deadtime by the factor

$N^{\rm GW}_{\rm vetoed}/N^{\rm aux}_{\rm coinc}$, so comparing the two ratios gives a rough idea as to what extent glitches are clustered. Even though UPV does not use these two ratios directly, they provide a useful means for sanity check and comparison with other vetoes.

Once all the figures of merit are calculated at each threshold, the UPV code creates veto segments for channels that pass the criteria described above. Using the KW significance threshold at which the channel exceeds a 50% used percentage, the veto segments are defined by placing a time window ($\pm$1s) around the peak times of auxiliary channel triggers above the threshold. To follow convention, all the veto segment times are rounded (widened only) to integer second values. For the inspiral searches, the veto segments are often further padded with extra time windows, because large SNR glitches tend to trigger various templates and produce cascade of triggers that could last longer than our original veto window [7, 8].

The UPV code is written in Python, using libraries that the LSC and Virgo have developed to manipulate the data. To avoid memory problems due to a large number of triggers, the code uses the SQLite database and works on a local disk, instead of manipulating data on memory. Also, some computationally heavy calculations are done on C-based libraries for better performance. The code is run on the LIGO data grid computers [9].

The code defines vetoes online for S6/VSR2 on a weekly basis. The veto results are used to improve the search for GWs emitted from compact binary coalescence systems [10]. Fig. 3 and Table 1 shows UPV performance on single detector inspiral triggers for the first 3 months of S6/VSR2 for each LIGO and Virgo detector (the 4 km interferometer at Hanford is called H1, the 4 km interferometer at Livingston is called L1, and the 3 km Virgo interferometer is called V1). We see that UPV is helping to clean up high SNR "outliers".

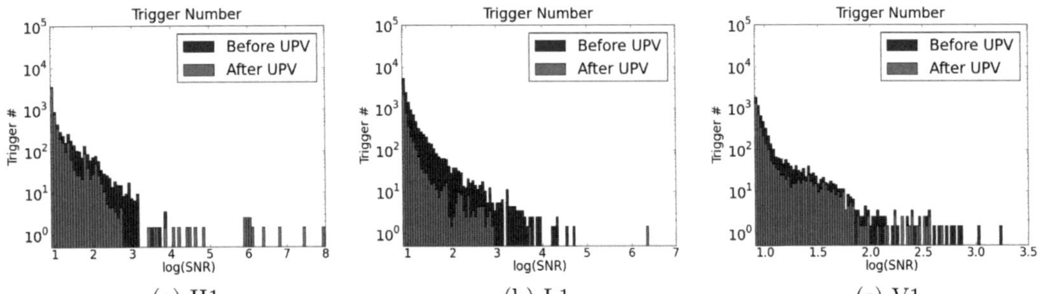

(a) H1 (b) L1 (c) V1

Figure 3: Histograms of single detector inspiral triggers for the first 3 months of S6/VSR2 before and after UPV is applied. Each histogram corresponds to the detectors called H1, L1 and V1 respectively. We can see that UPV removes many "outliers".

The search for GW bursts uses a related veto method called "hveto" that searches for statistically significant auxiliary channels [11]. Even though hveto uses a different ranking statistics, the results agree well with UPV, providing assurance for our method.

	H1	L1	V1
Deadtime	0.604%	1.062%	0.475%
Efficiency, SNR> 8	23.0%	70.4%	42.1%
Efficiency, SNR> 10	42.4%	79.8%	51.6%
Efficiency, SNR> 20	58.2%	85.9%	40.1%

Table 1: Efficiencies of UPV on single inspiral triggers for the first 3 months of S6/VSR2.

2.1. Safety

Since the UPV is based on statistical measurements without regards to causality, we need to ensure that the auxiliary channels have negligible sensitivity to GWs. In particular, GW signals could couple into some auxiliary channels, leading us to veto GW signals rather than noise by finding coincidences between the GW channel and the auxiliary channels. Auxiliary channels with a non-negligible coupling from GW signals are called "unsafe". Some channels are already identified as unsafe from the past analysis and input from the commissioning team. We disregard these as veto channels *a priori*. (There are, however, some suggested veto methods to make use of such unsafe channels [13].)

Since the detectors are highly complicated, there is always a possibility of unknown coupling mechanisms. Therefore, we take another safety precaution using simulated GW signals inserted into the data by physically shaking the mirrors (hardware injections). The number of false dismissals of hardware injections should be consistent with an accidental coincident rate calculated from the deadtime, and with a sufficient number of injections, we can establish the safety of vetoes. We calculate the probability that the channel would veto at least N injections via accidental coincidence:

$$\text{Safety Probability} \equiv 1 - F(N^{\text{inj}}_{\text{vetoed}} - 1; N^{\text{inj}}_{\text{exp}}) \tag{7}$$

where F is Poisson cumulative density function, $N^{\text{inj}}_{\text{vetoed}}$ is the number of injections actually vetoed, and $N^{\text{inj}}_{\text{exp}}$ is the expected number of injections to be vetoed from the deadtime assuming triggers are randomly distributed. We do not use channels that have a safety probability smaller than 10^{-5}. This value was chosen empirically; the past analysis has shown that statistical fluctuations would not give probability lower than 10^{-3} for channels we believe safe, while known unsafe channels usually show up with probability $< 10^{-8}$. Finally, all the veto candidate channels are reviewed by the LSC's detector characterization group and glitch group [12] for safety before being implemented as vetoes; Virgo vetoes are similarly reviewed by the Virgo data quality group.

3. Application to Detector Characterization

Even though the UPV was initially developed for defining vetoes, it can be a useful diagnostic tool for the detectors. For example, while we were following up the outcome of the UPV for the last science run (S5), we identified that some noise was due to sledge hammers used in construction work near the observatory. Motivated by this example, we further developed the code so that we could apply the UPV for detector characterization in a more systematic way. To apply the output of the UPV for detector characterization, we tuned some of the criteria differently from defining vetoes; we required a 30% used percentage instead of 50% and also required only 5 $N^{\mathrm{aux}}_{\mathrm{coinc}}$ in a day instead of 10 in a week, in order to allow more channels to be analyzed. We developed a postprocessing code that gathers all the coincident triggers from veto candidate channels, clusters the events with 1s window, reorders them according to the corresponding significance of the GW trigger, and creates a webpage for further human investigations for these events that have at least two channels in coincidence with the GW channel. The idea is to classify noise by corresponding channels, and analyze each noise group to identify the physical source. The code uses information from "data quality (DQ) flags" [4, 5, 7] and launches "Omega Scans" (formerly known as "Q Scans" [6]). DQ flags are defined and stored in a database to provide information on data quality. They mark noise with well-known origins, such as seismic noise from train or wind, as well as the state of the detectors, and thus provide useful information for the investigation. Omega Scan is a tool to perform a detailed study of the LIGO and Virgo data stream around a specific time of interest. In addition to displaying the time series and time-frequency spectrograms of the GW channel data, Omega Scan can also efficiently search a large number of auxiliary channels for statistically significant signal content (see Fig. 4 for an example of the Omega Scan). Both DQ flags and Omega Scans are linked to the webpage that the postprocessing code creates. The basic work flow of the UPV is shown in Fig. 5. The UPV and the postprocessing codes are launched every day, and the results are posted for further investigation.

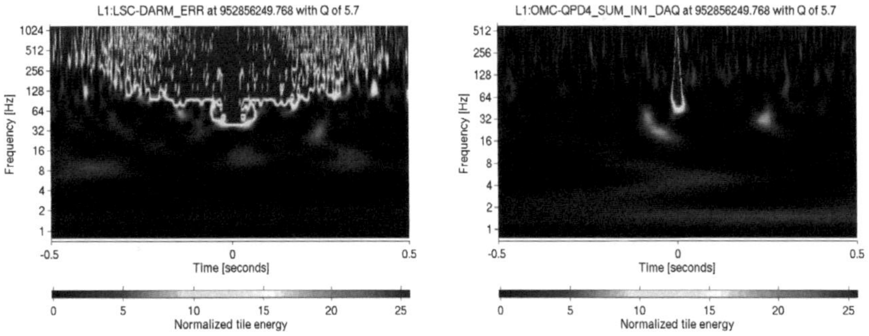

(a) GW channel (b) Auxiliary channel

Figure 4: Examples of Omega Scans [6]. These are time-frequency representation of a particular glitch in the GW channel (a) and an auxiliary channel (b). UPV removed this kind of glitch using an output mode cleaner channel. We are using information obtained through the UPV process to track down the actual cause of glitches.

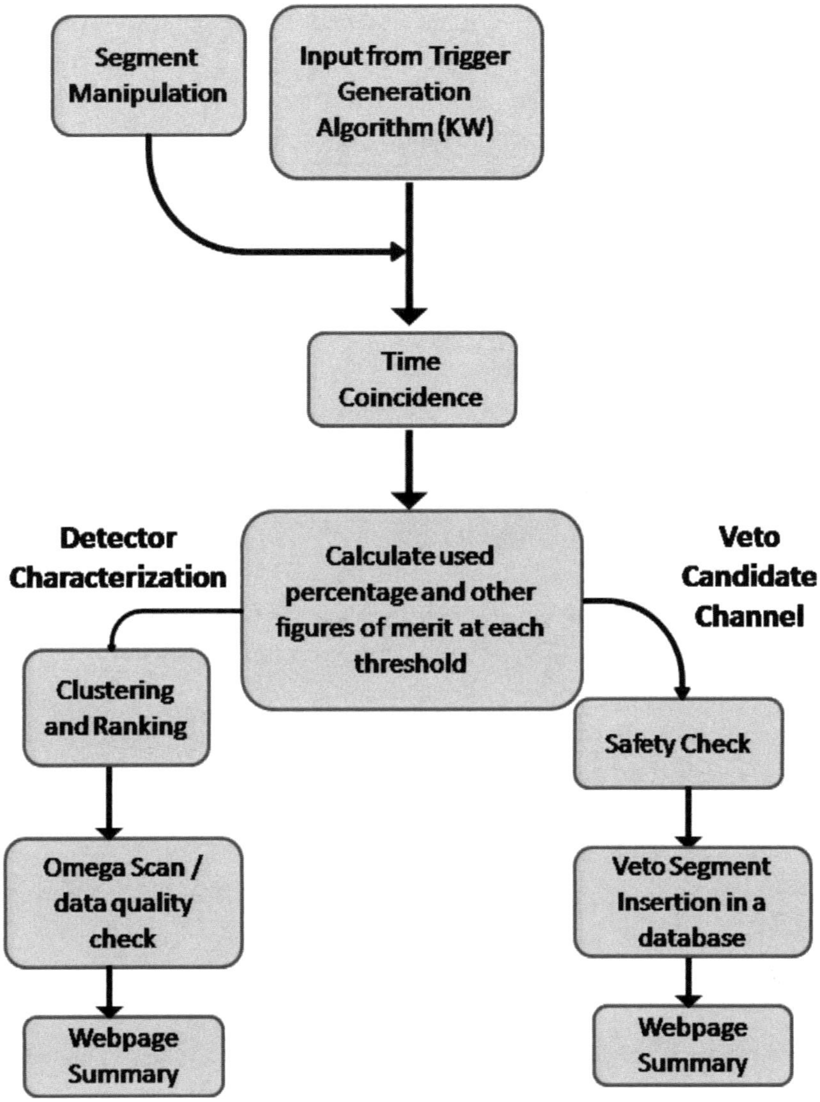

Figure 5: A diagram showing the workflow of UPV for a single auxiliary channel. Figures of merit including used percentage are calculated at each threshold after particularly noisy segments have been excluded. We define vetoes for channels with a strong correlation with the GW channel. In addition, we use information obtained from this process for detector characterization. This process is done on over 150 auxiliary channels for each of LIGO detectors and over 500 channels for the Virgo detector.

4. Summary

The UPV is contributing to the elimination of transients in LIGO and Virgo data that are difficult to remove by other methods. For the ongoing S6 and VSR2 science run, the

UPV is running week by week automatically, defining vetoes for binary inspiral searches, and helping to identify and eliminate a large number of problematic noise triggers.

In addition, the UPV is running every day for detector characterization. It is monitoring day-to-day changes of the detectors with respect to their auxiliary channels, and providing quick feedback about the malfunctioning of some components of the detectors. It also supplies useful information on identifying the physical origin of noises, thereby contributing to the improvement of the detectors.

5. Acknowledgments

This project is funded by NSF Grant PHY-0854790, Summer Undergraduate Research Fellowship and the Kolenkow-Reitz Fund at Carleton College. The authors gratefully acknowledge the support of the United States National Science Foundation for the construction and operation of the LIGO Laboratory, the Science and Technology Facilities Council of the United Kingdom, the Max-Planck-Society, and the State of Niedersachsen/Germany for support of the construction and operation of the GEO600 detector, and the Italian Istituto Nazionale di Fisica Nucleare and the French Centre National de la Recherche Scientifique for the construction and operation of the Virgo detector. The authors also gratefully acknowledge the support of the research by these agencies and by the Australian Research Council, the Council of Scientific and Industrial Research of India, the Istituto Nazionale di Fisica Nucleare of Italy, the Spanish Ministerio de Educacion y Ciencia, the Conselleria dEconomia Hisenda i Innovacio of the Govern de les Illes Balears, the Foundation for Fundamental Research on Matter supported by the Netherlands Organisation for Scientific Research, the Polish Ministry of Science and Higher Education, the FOCUS Programme of Foundation for Polish Science, the Royal Society, the Scottish Funding Council, the Scottish Universities Physics Alliance, The National Aeronautics and Space Administration, the Carnegie Trust, the Leverhulme Trust, the David and Lucile Packard Foundation, the Research Corporation, and the Alfred P. Sloan Foundation.

Reference

[1] Abbott B *et al* (the LIGO Scientific Collaboration) 2009 LIGO: the Laser Interferometer Gravitational-Wave Observatory *Reports on Progress in Physics* **72** 076901

[2] Acernese F *et al* (the Virgo Collaboration) 2009 Status of Virgo *Class. Quantum Grav.* **25** 114045

[3] Grote H (the LIGO Scientific Collaboration) 2008 The Status of GEO 600 *Class. Quantum Grav.* **25** 114043

[4] Christensen N (for the LIGO Scientific Collaboration and the Virgo Collaboration) 2010 LIGO S6 Detector Characterization Studies submitted to these GWDAW14 proceedings, LIGO Document P1000045

[5] Robinet F (for the LIGO Scientic Collaboration and the Virgo Collaboration) 2010 Data Quality in Gravitational Wave Burst and Inspiral Searches in the Second Virgo Science Run submitted to these GWDAW14 proceedings, LIGO Document P1000047

[6] Chatterji S, Blackburn L, Martin G and E. Katsavounidis E 2004 Multiresolution techniques for the detection of gravitational-wave bursts *Class. Quantum Grav.* **21** S1809

[7] Slutsky J *et al* 2010 Methods for Reducing False Alarms in Searches for Compact Binary Coalescences in LIGO Data *arXiv:1004.0998*

[8] Christensen N, Shawhan P and Gonzalez G (for the LIGO Scientific Collaboration) 2004 *Class. Quantum Grav.* **21** S1747-S1755

[9] `https://www.lsc-group.phys.uwm.edu/lscdatagrid/overview.html`

[10] Abbott B *et al* (the LIGO Science Collaboration) 2009 Search for gravitational waves from low mass binary coalescences in the first year of LIGO psila s S5 data *Phys. Rev. D* **79** 122001

[11] Abbott B *et al* (the LIGO Science Collaboration) 2009 Search for gravitational-wave bursts in the first year of the fifth LIGO science run Phys. Rev. D **80**, 102001

[12] Blackburn L *et al* 2008 The LSC glitch group: monitoring noise transients during the fifth LIGO science run *Class. Quantum Grav.* **25** 184004

[13] Hild S *et al* 2007 A statistical veto method employing an amplitude consistency check *Class. Quantum Grav.* **24** 3783-3798

Journal of Physics: Conference Series **243** (2010) 012006 doi:10.1088/1742-6596/243/1/012006

Classification of glitch waveforms in gravitational wave detector characterization

S. Mukherjee[1], R. Obaid, B. Matkarimov[2]
Center for Gravitational Wave Astronomy
University of Texas at Brownsville, Brownsville, USA

[1] Corresponding author: soma@phys.utb.edu
[2] Permanent address: Kazakh-British Technical University, Kazakhstan

PACS numbers: 95.55.Ym, 95.75.Wx, 95.75.Pq

Abstract

The paper describes a new fast and efficient algorithm to classify waveforms from signals that arise in gravitational wave detectors like LIGO. Such waveform classification is useful and important for detector characterization as well as for understanding glitches seen in the analysis pipelines that detect signals from astrophysical burst and compact object inspiral sources. Classification of glitches based on discrete parameters has been reported earlier by the author. In the current study, a new feature-mining approach has been developed that uses the additional information based on shape of the glitch waveforms. This has been a challenging problem because of the unique structures present in the real gravitational wave data. The paper presents results from simulations as well as real data. Studies show that the proposed method is fast and efficient in classifying glitches in noise.

PACS numbers: 95.55.Ym, 95.75.Wx, 95.75.Pq

1. Introduction

The three LIGO detectors at Hanford and Livingston [1] have been taking data in 'science mode' since July 2009. This is the sixth science run (S6) of LIGO [2]. Raw data from the gravitational wave (GW) interferometers can often be non-stationary, containing narrow band noise, as well as broadband glitches (also known in literature as transients or triggers) [3]. Presence of glitches can severely restrict the efficiency of astrophysical searches and hence we need to understand the origin of these glitches – instrumental, environmental or otherwise. The glitches themselves (often arriving at a high rate) may mimic gravitational waves. It is thus of utmost importance to develop data analysis methods that can mine the glitch database to extract information about the origin and properties of the glitches. There is a two-fold advantage in this method. First, building up a knowledge base for the different types of observed glitches can help model the GW noise accurately and realistically for refinement of GW detection statistic. Secondly, tracing their instrumental origin can help experimentalists with debugging noise.

There are several pipelines operating online on LIGO data viz. kleine welle (KW) [4] and waveburst (WB) [5]. The KW pipeline, which works on the GW channel and several hundreds of auxiliary and environmental channels, picks up glitches at a fairly high rate. Unsupervised data mining methods like the multidimensional hierarchical classification analysis [6, 7] have been developed and LIGO science data have been analyzed in recent past. The aim of these studies has been to classify the population of glitches seen in the GW, environmental and auxiliary channels into statistically significant distinct groups with uniform characteristics. These methods use discrete properties of the glitches and produce groups in the higher dimensional parameter space, effectively reducing the dimensionality of the problem. However, in these analyses methods, we do not incorporate an important aspect of the glitches.

Glitches in GW data contain very rich information in terms of *shape* or *glitch waveform*. One can also make use of this information content in the time series data itself in classifying the glitches. This brings us to the aspect of development and application of methods in temporal classification [8]. Temporal classification methods e.g. S-means and Constraint Validation have been developed and studied on simulated GW data [9,10]. While these methods showed very promising results on simulated waveforms, they have not been tested on data from real GW detectors. Moreover, S-means has been built around the Dynamic Time Warping [11] which may not be optimal for heterogeneous datasets as seen in GW data from detectors like LIGO and Virgo [12]. Dynamic programming can also be slow quite often since it involves an element-by-element comparison of the time series. While his problem can be countered by developing index-based methods [10, 13-15], it is not totally free from problems because these methods may require using dynamic programming on the superset.

The current paper describes a new fast and efficient method for unsupervised classification of glitch waveforms. This method is based on distances calculated using Longest Common Subsequence (LCSS) [16]. The technique is studied in GW data for the first time. Results are demonstrated using, (i) noise-free simulated waveforms and (ii) waveforms embedded in Gaussian white noise. After benchmarking, the method has also been applied to KW database from LIGO's fourth science run (S4).

2. Kleine Welle algorithm

Kleine Welle (KW) is an algorithm uses a dyadic Haar wavelet [17] transform to look for excess energy in time scale decompositions. The statistical outliers are identified by setting a threshold on the individual pixel energy.The significance of the nearby pixels that cluster on the time–frequency plane is calculated from known distributions.

3. Glitch waveform classification based on LCSS based algorithm

There are two main issues that one must address while developing an efficient feature-based classification method – (i) the data may have temporal gaps, i.e. there may be same pattern occurring at different time epochs and (ii) the classification preferably should be automatic (to keep up with the online feedback systems) and thus unsupervised classification methods *robust to noise* need to be explored.

Let us first take a look at why LCSS is efficient and how it fits into GW data analysis. In any unsupervised classification method, the first step is to calculate the distance between points in the parameter space. Even though Euclidean distance is the most commonly used method, it is not suitable for addressing the first difficulty mentioned above. Thus, two glitches that have the same waveform may show a high Euclidean distance if they are not occurring simultaneously. This will be considered a redundant cluster from the physical point of view. LCSS is able to compute a match between two time series by calculating distances that do not necessarily occur at the same time *without having to rearrange the sample sequence.*

Let X and Y be two waveforms with size p and q respectively.

$$X = [(x_{i,1}, x_{j,1}), \ldots, (x_{i,p}, x_{j,q})] \quad (1)$$
$$Y = [(y_{i,1}, y_{j,1}), \ldots, (y_{i,p}, y_{j,q})] \quad (2)$$

Let us define $\Delta X = [(x_{i,1}, x_{j,1}), \ldots, (x_{i,p-1}, x_{j,q-1})]$ (3)

Given an integer d and a number e ($0<e<1$), we define the $LCSS_{d,e}(X, Y)$ as

$LCSS_{d,e}(X, Y) = 0$ *if* X or Y is empty sequence (4i)

Journal of Physics: Conference Series **243** (2010) 012006 doi:10.1088/1742-6596/243/1/012006

$$\mathrm{LCSS}_{d,e}(X, Y) = 1 + \mathrm{LCSS}_{d,e}(\Delta X, \Delta Y) \; \textit{if} \; | x_{i,p}, y_{j,q}| < e \; \textit{and} \; | x_{i,p}, y_{j,q}| < e \; \textit{and} \; | p - q| <= d \qquad (4ii)$$

$$\mathrm{LCSS}_{d,e}(X, Y) = \max(\mathrm{LCSS}_{d,e}(\Delta X, Y), \mathrm{LCSS}_{d,e}(X, \Delta Y) \; \textit{otherwise}. \qquad (4iii)$$

d denotes a temporal match while and e denotes the threshold in space.

The distances arrived at from the LCSS can be defined as follows –

$$\Gamma_{d,e}(X, Y) = 1 - \mathrm{LCSS}_{d,e}(X, Y) / \min(p,q) \qquad (5)$$

Once the distances are calculated according to equation (5), a generalized k-means [29] algorithm is employed to form the clusters. The reason as to the choice of k-means is dictated by the fact that this allows formation of homogeneous clusters that are insensitive to outliers. Generalized k-means [30] uses two parameters to start with – the number of clusters m and an initial seed s. The algorithm works as follows.

Let T be a set of M waveforms. $T = \{T_1, \ldots, T_M\}$. Let the number of clusters be m. Let there be a set of m seeds denoted by $S = \{s_1, \ldots s_m\}$. Each seed s_j is composed of s time series representation of the waveforms, chosen randomly in T. The choice of seed can be random in absence of *a priori* knowledge about the dataset, as is often the case with unsupervised clustering. The distance between the waveform T_i and the seed s_j is given by

$$\Lambda(T_i, s_j) = 1/s \; \Sigma_{z \,\varepsilon\, sj} \; \mathrm{LCSS}(T_i, z). \qquad (6)$$

Let G be the set of m clusters, where $G = \{G_1, \ldots, G_m\}$. Each cluster G_j is made up of all waveforms that have s_j as nearest seed. He next step is to obtain a new set of seeds $S^* = \{s^*_1, \ldots s^*_m\}$ where each new s^*_j is obtained by minimizing a function

$$I(T_i, s_j) = \Sigma_{z \,\varepsilon\, Gj} \; \mathrm{LCSS}(T_i, z). \qquad (7)$$

$I(T_i, s_j)$ is usually called the *inertia function.* In the final step we calculate the intra-class variance

$$\sigma^2(G) = 1/M \; \Sigma^{k}_{j=1} \; \Sigma_{Ti \,\varepsilon\, Gj} \; I(T_i, s_j). \qquad (8)$$

The iterative process is continued until the value of σ^2 (G) does not fall below a given threshold.

4. Analysis Results

We carry out the analysis in three steps. First we implement the algorithm on different types of simulated signals. In the second phase we analyze simulated signals of various forms embedded in noise and in the third and final phase we run the algorithm on LIGO's fourth science run data (S4).

a. Simulated waveforms without noise

We first generate a data set of 350 simulated waveforms with variable parameters, each 1024 samples long. The waveforms are in the shapes of sine Gaussians, Gaussian pulses, Chirps and Mixture sinusoids with varying parameters, i.e. varying amplitude, frequency, width, location on the time axis etc. The waveform database is introduced to the pipeline as the prime input. Figure 1 summarizes the output of the classification. The algorithm correctly classified the waveforms into the right categories. No misclassification was noted. This is an expected result in the absence of noise.

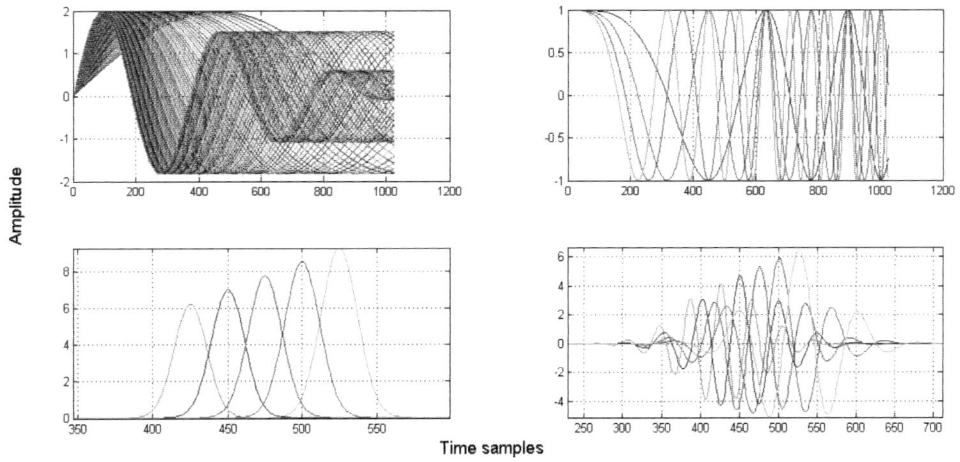

Figure 1. This figure shows the result of classifying 350 simulated waveforms. Each subplot shows classes of waveforms as selected by the LCSS classification scheme. The input waveforms came in four different shapes - sine Gaussians (bottom right), Gaussian pulses (bottom left), Chirps (top right) and Mixture Sinusoids (top left) with varying parameters. The x-axis gives the time samples and the y-axis gives amplitude measured in arbitrary units.

b. Simulated waveforms embedded in noise

In the next step of this study, we generate a data set of 350 simulated waveforms of four different types of shape with variable parameters, each 1024 samples long, as described in the previous section. Each waveform data stream is mixed with Gaussian white noise. The output is thus a noisy waveform. This is fed to the classification pipeline. The results are summarized in Figure 2. In this case too, the LCSS algorithm could classify the waveforms with 100% accuracy. This *robust-against-noise* result is rather optimistic and points towards its aptness to be used in classification of glitches noted in GW data. It was interesting to observe that, when the same data was analyzed using standard k-means algorithm (without invoking the LCSS), approximately 23% of the glitches were misclassified.

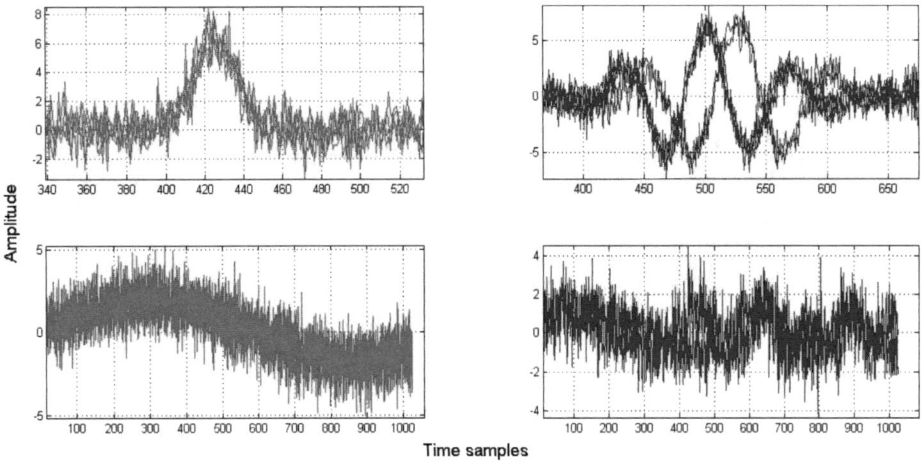

Figure 2. This figure shows the result of classifying 350 simulated waveforms in presence of Gaussian white noise. Each subplot shows classes of waveforms as selected by the LCSS classification scheme. The input waveforms came in four different shapes - sine Gaussians (top right), Gaussian pulses (top left), Chirps (bottom right) and Mixture Sinusoids (bottom left) with varying parameters. The x-axis gives the time samples and the y-axis gives amplitude measured in arbitrary units.

c. Glitches in LIGO science data

Having gained insight with the simulations, we now apply the LCSS algorithm to classify glitches seen in LIGO science data. We have used glitches found in the KW glitch catalog during LIGO's fourth science run (S4) [18]. The complete pipeline is explained in figure 3. This involves two main stages – (i) Extraction of the glitch from noisy output of the gravitational wave channel and (ii) Analysis of the extracted S4 glitch waveforms by the LCSS algorithm. Stage (i) involves the following sequence.

- (i) Selection of glitches with signal-to-noise ratio (snr) greater than 10 from KW database;
- (ii) Extraction of raw GW channel data, centered around the glitch (extracted time series noted by, say, $\{q_i\}$.)
- (iii) Whitening $\{q_i\}$ [19] and dynamically removing [20] the narrowband noise present in $\{q_i\}$. The resulting time series is denoted by $\{cq_i\}$.
- (iv) Filtering $\{cq_i\}$ with a bandwidth of 256 Hz around f_c, where f_c is the central frequency of the glitch as noted in the KW catalog. The resulting time series is denoted by $\{fcq_i\}$.
- (v) Heterodyning $\{fcq_i\}$ and re-sampling the resulting data.

All the extracted glitches are normalized and stored as a time series of 1024 samples. This is introduced as the input to the LCSS pipeline for classification of the glitches.

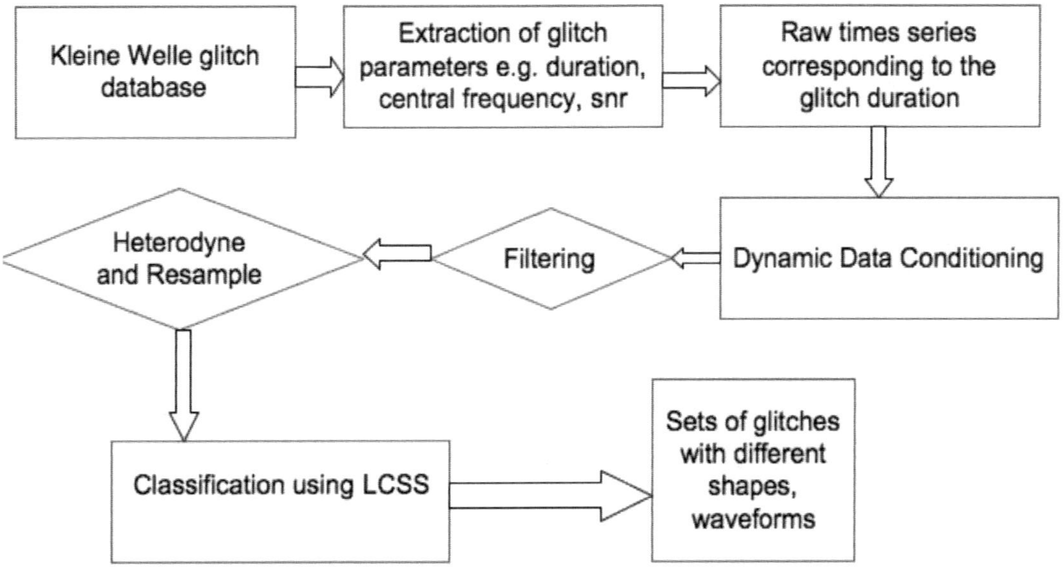

Figure 3. This is the schematic diagram of the LCSS pipeline for classification of glitches seen in the data from present generation of GW interferometers. A full explanation of the pipeline is given in section 4c.

In the data set used (an arbitrary day in S4), we extracted 150 glitch waveforms using the method described above. Figure 4 shows some examples of the typical glitch shapes noticed in the population under study. The examples amply demonstrate that there is a wide repertoire of glitch shapes in the data and thus merits classification. Figure 5 shows the end results of classification pipeline. Eight distinct and statistically significant classes of glitches were detected. It was found that the representative groups (classes) were consistent in picking up glitches of similar shapes or waveforms.

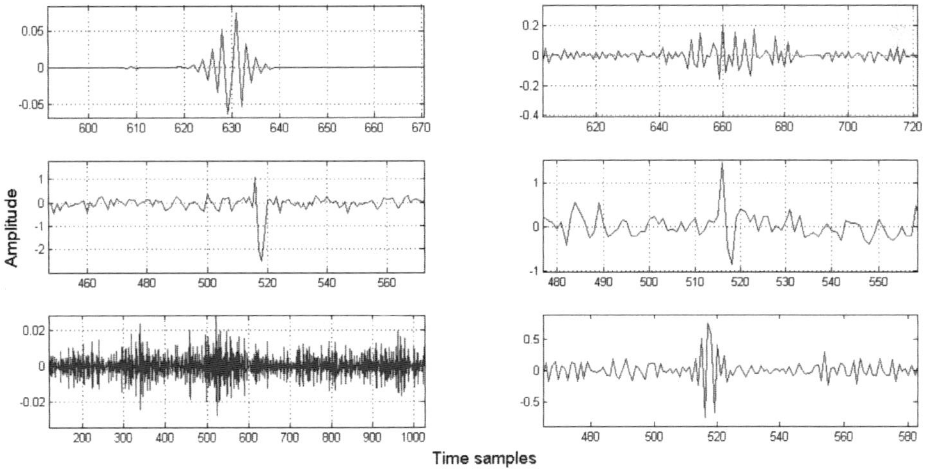

Figure 4. This figure shows examples of some of the typical types of glitch shapes seen in the population. The x-axis gives the time samples and the y-axis gives amplitude measured in arbitrary units.

Journal of Physics: Conference Series **243** (2010) 012006 doi:10.1088/1742-6596/243/1/012006

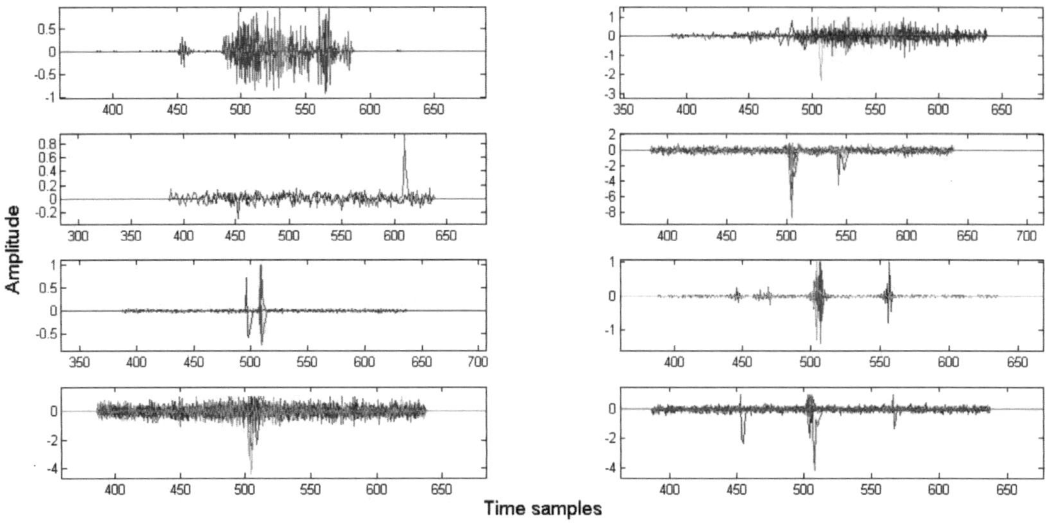

Figure 5. The figure shows eight distinct and statistically significant classes of glitches as detected by the LCSS pipeline on a population of glitches seen in an arbitrary day during LIGO's fourth science run. The x-axis gives the time samples and the y-axis gives amplitude measured in arbitrary units.

5. Discussion and future direction

Once the classes of glitches are determined, it becomes imperative to subject them to follow-up studies to determine their possible sources, whether instrumental or environmental. There exists a rich collection of various methods within the LIGO Scientific Collaboration (**www.ligo.org**) that helps in this direction. Methods such as Omega Scan [21], H-veto [22], detector-wide channel scan (DWCS) [23] or online KW analysis [24] can be very helpful in following up on the classified glitch groups. There also exists a large repository of data quality (DQ) flags (25-28). These can also serve as complementary information in conjunction with the follow-up methods mentioned above. The examples in this paper are shown on S4 data. Since then, a huge development has taken place in these areas of detector characterization and the current search tools and glitch characterization machineries can indeed bring out valuable information regarding explanation of observed glitches, thus providing a better handle to the detector noise. For example, one could take the possible set of correlated auxiliary and environmental channel glitches in the same time window (following the schemes given by say, H-veto or DWCS) and check the waveforms of the coincident (within the accuracy of the same time window) glitches for consistency. It would be an interesting study to see if a set of waveforms of similar shape could be correlated to a particular subset of auxiliary channels. One would thus develop a complete repository of glitch waveforms that could be used as a discriminator in detector characterization studies as well as in astrophysical search engines where either the waveform from a source could be known (coalescence of compact objects) or unknown (as in the case of burst sources). This is not done yet and can, in principle, open up a new window of understanding glitch sources in GW data. This is under construction by the authors. The plan is to run the complete analysis on the current S6 data.

To summarize, we have developed and tested a new fast and efficient method based on distances calculated using Longest Common Subsequence (LCSS) for unsupervised classification of glitch waveforms. The technique is studied in GW data for the first time. The method proves promising for classification of glitches in presence of noise because of its robust character. The method is now being applied on S6 data.

Acknowledgment

The authors gratefully acknowledge the LIGO Laboratory for use of LIGO S4 data. This work is supported by NSF grant #PHY0855371, 2009. The document is assigned LIGO Technical document number P1000057.

References

1. Abramovici A *et al* 1992 *Science* **256** 325–33

2. Kawabe, K. 2008 *J. Phys.: Conf. Ser.* **120** 032003

3. Mukherjee S 2003 *Class. Quantum Grav.* **20** 925–36

4. Chatterjee,S. et al. *Classical and Quantum Gravity*, **21**, S1809-18, 2004

5. Klimenko, S *et al* 2004 *Class. Quantum Grav.* **21** S1685

6. Mukherjee S 2006 *Class.Quantum Grav.* **23** S661–71

7. Mukherjee S (for the LSC) 2007 *Class.Quantum Grav.* **24** S701

8. Schoenfeld, W. N. and Cumming, W. W. 1960 *Proceedings of the National Academy of Sciences of the United States of America* **46** 753-758

9. Tang L R et al. 2008 *Class. Quantum Grav.* **25** 184023

10. Lei, H *et al* 2007 *International Workshop on Knowledge Discovery from Ubiquitous Data Streams, 18th European conference on Machine Learning (ECML) and 11th European conference on Principles and Practice of Knowledge Discovery in Databases (PKDD)*, Warsaw, Poland. ECML, PKDD proceedings, Springer-Verlag (LNAI series)

11.Sakoe, H and Chiba, S 1978 *IEEE Trans. Acoustics, Speech and Signal Processing*, ASSP, **26**, 43-49

12. Cavalier, F 1999 *34th Rencontres de Moriond : Gravitational Waves and Experimental Gravity*, Les Arcs, France

13. Agarwal, P. K. *et al* 2000 19th *ACM symposium on Principles of database systems (PODS)*

14. Pfoser, D 2000 *VLDB,* Cairo

15. Saltenis, S et al 2000 *ACM SIGMOD*

16. Vlachos, M 2002 *ICDE* 673-84

17. Mallat, S 1999 A *Wavelet Tour of Signal Processing (Wavelet Analysis and its Applications)* 2nd edn (New York: Academic)

18. Credico, A 2005 *LIGO technical document* G050292-00-Z

19. Mohanty, S 2002 *Class. Quantum Grav* **19** 1513

20. Weerathunga, T. S. 2010 *M.S. Thesis, University of Texas at Brownsville*

21. Sutton, P. et al 2009 *gr-qc* 0908.3665

22. Smith, J (for the LSC) 2008 *Class Quantum Grav* **26** 114013

23. Mukherjee, S 2007 *GWDAW12*, Boston, *LIGO technical document* G070814-00-Z

24. Christensen, N 2010 *Class. Quantum Grav* This issue

25. Gonzalez G *et al* 2007 *LIGO technical document* G070142-00-Z

26. Blackburn L *et al* 2006 *LIGO technical document* G-060407-00-Z

27. Chatterjee S 2007 *LIGO technical document* G-070138-00-Z

28. Blackburn L 2005 *LIGO technical document* G-050158-00-Z

29. Diday, E 1973 *International Journal of Computer Sciences* **2**

30. McQueen, J 1967 *5th Berkeley Symposium on Mathematical Statistics and Probability*, Berkeley, CA

Complete phenomenological gravitational waveforms from spinning coalescing binaries

R. Sturani(1,2), **S. Fischetti**(3), **L. Cadonati**(3), **G. M. Guidi**(1,2), **J. Healy**(4), **D. Shoemaker**(4), **A. Viceré**(1,2)

(1) Dipartimento di Matematica, Fisica e Informatica, Università degli Studi di Urbino 'Carlo Bo', I-61029 Urbino, Italy
(2) INFN Sezione di Firenze, I-50019 Sesto Fiorentino, Italy
(3) Physics Department, University of Massachusetts, Amherst MA 01003, USA
(4) Center for Relativistic Astrophysics, Georgia Tech, Atlanta, GA 30332, USA

E-mail: riccardo.sturani@uniurb.it

Abstract. The quest for gravitational waves from coalescing binaries is customarily performed by the LIGO-Virgo collaboration via matched filtering, which requires a detailed knowledge of the signal. Complete analytical coalescence waveforms are currently available only for the non-precessing binary systems. In this paper we introduce complete phenomenological waveforms for the dominant quadrupolar mode of generically spinning systems. These waveforms are constructed by bridging the gap between the analytically known inspiral phase, described by spin Taylor (T4) approximants in the restricted waveform approximation, and the ring-down phase through a phenomenological intermediate phase, calibrated by comparison with specific, numerically generated waveforms, describing equal mass systems with dimension-less spin magnitudes equal to 0.6. The overlap integral between numerical and phenomenological waveforms ranges between 0.95 and 0.99.

1. Introduction

The Laser Interferometer Gravitational-wave Observatory (LIGO) and Virgo constitute a network of kilometer-scale interferometers for the detection of gravitational waves. The initial detector configuration has successfully acquired data at design sensitivity; substantial upgrades over the next few years, to Advanced LIGO and Advanced Virgo, are expected to yield the detection of gravitational waves from the coalescence of compact binary systems of neutron stars or black holes.

The coalescence of binary systems is usually described in terms of three distinct phases: the inspiral, the merger and the ring-down. The inspiral phase allows for an accurate analytical description via the so-called Post-Newtonian (PN) expansion, see for instance [1] for a review. The ring-down also admits a perturbative analytical model, as it describes the damped oscillations of the single object resulting from the binary coalescence, as a superposition of black-hole quasi normal modes [2]. The merger phase is however fully non-perturbative and for generic systems it has not been described analytically but rather by numerical simulations. During the last five years numerical relativity has made tremendous progress in describing the full coalescence of a binary system beginning with [3, 4, 5] and more recently [6, 7, 8, 9, 10, 11], see [12, 13, 14] for reviews and it can now produce waveforms for generic spin orientations, with moderate spin magnitude ($\lesssim 0.9$) and mass ratios ($\lesssim 10:1$).

Matched filtering is typically used in LIGO-Virgo data analysis in order to uncover weak signals buried into noise [15]. This technique requires a large number (tens of thousands) of templates to be checked against real data. Due to the computational cost of numerical simulations, this is only possible with an analytical knowledge of the templates. Analytical methods are currently available to reproduce the complete waveform emitted by non-precessing coalescing binaries. This has been achieved in the Effective One Body (EOB) construction [16, 17, 18] for non-spinning systems, and in the EOB-spin waveforms [19], an extension of the EOB method to the case of binaries with non-vanishing spins aligned with the orbital angular momentum. Both spinning and non spinning EOB require a comparison with numerically generated waveforms in order to calibrate some free parameters of the model. Another method for generating analytical waveforms for spinning non-precessing binaries is by joining PN-generated inspiral and numerical waveforms [10, 20] to construct phenomenological waveforms.

In this paper we present a new family of analytical waveforms from the coalescence of generic spinning binaries.

The waveforms are obtained by interpolating between the perturbative PN inspiral description and the ring-down, both admitting physically motivated analytic models, with a phenomenological phase over the merger portion of the signal. Since in the matched-filtering search an accurate determination of the gravitational phase of the signal is crucial, more important than an accurate amplitude determination, we focus on the phenomenological parametrization of the gravitational wave phase, and give lower priority to the amplitude. Spinning waveforms depend on several parameters (masses, spin components of the binary constituents, angles defining the orientation of the source with respect to the observer). For this reason it is not practical to use these waveforms to build template banks, but we envision their usefulness as injection waveforms in testing existing LIGO-Virgo data analysis pipelines, which are currently based on non-spinning templates.

The paper is organized as follows. In sec. 2 the method for the analytical waveform construction and the numerical simulations are illustrated. In sec. 3 the results are presented, in the form of comparison between analytically and numerically generated waveforms. We restricted the analysis to the main quadrupolar mode ($m = 2$), but this setup can be extended to include other modes. In sec. 4 the conclusions that can be drawn from the present work are reported.

2. The method

The goal of the present work is to produce the complete analytical waveforms generated by the coalescence of spinning binary systems. The numerical waveforms used to construct and calibrate our analytical model all describe equal mass binary systems ($m_1 = m_2$), with spin magnitudes $|\mathbf{S_1}| = |\mathbf{S_2}| = 0.6\, m_1^2$ and $\mathbf{S_2}$ orthogonal to the initial orbital angular momentum (where $\mathbf{S_{1,2}}$ denote the binary constituent spin vectors and we posit $G_N = c = 1$).

The description of the dynamics adopted here models the inspiral phase via the standard TaylorT4 PN formulas, see [21] for definition and comparison of different PN approximants in the spin-less case. In the non-spinning case the waveform is fully parametrized by amplitude and the orbital phase, but in the spinning case the effect of the precession of the orbital plane should also be accounted for, see [22] spin and angular momentum evolution equations. It is convenient to define an orbital phase $\phi = \int \omega_{orb}\, dt$ whose evolution is given by

$$\frac{d\phi}{dt} = \frac{v^3}{M}, \qquad \frac{dv}{dt} = -\frac{F(v)}{dE/dv}, \tag{1}$$

where $M \equiv m_1 + m_2$ is the total mass of the binary system, $F(v)$ and $E(v)$ are respectively the

Journal of Physics: Conference Series **243** (2010) 012007 doi:10.1088/1742-6596/243/1/012007

flux emitted and the energy of a circular orbit with angular frequency $\omega_{orb} = v^3/M$. The main gravitational wave frequency f_{GW} is related to ω_{orb} via $f_{GW} = \omega_{orb}/\pi$.
In order to determine the actual waveform also the spins and the orbital angular momentum have to be dynamically evolved, see e.g. [23]. In particular by parametrizing the orbital angular momentum unit vector $\hat{\mathbf{L}}$ as $\hat{\mathbf{L}} = (\sin\iota\cos\alpha, \sin\iota\sin\alpha, \cos\iota)$ it is convenient to introduce the carrier phase Ψ given by

$$\frac{d\Psi}{dt} = \omega_{orb} - \cos\iota\,\frac{d\alpha}{dt}\,. \qquad (2)$$

Numerically generated waveforms are usually decomposed in spherical harmonics, in particular the five quadrupolar modes ($l = 2$) are the only non-vanishing at the lowest order in v, and the $l = 2, m = \pm 2$ mode are the dominant ones. The actual shape of the $l = 2, m = 2$ mode, the only one which will be used here for comparison with numerical relativity results, is given by the following [23]

$$h_{2,2}^{(insp)} = -2\frac{\nu M v^2}{d}\sqrt{\frac{16\pi}{5}}\left[\cos^4(\iota/2)\cos\left(2(\Psi+\alpha)\right) + \sin^4(\iota/2)\cos\left(2(\Psi-\alpha)\right) + O(v)\right]\,, \qquad (3)$$

where $\nu \equiv m_1 m_2/M^2$ is the symmetric mass ratio, d the source-observer distance and this formula refers to the inspiral waveform only.
The functions $F(v)$ and $E(v)$, necessary to determine the orbital phase, are known up to 3.5PN order as far as orbital effects are concerned, and up to 2.5PN and 2PN level for respectively $\mathbf{S_{1,2}L}$ and $\mathbf{S_1S_2}$, $\mathbf{S_1S_1}$, $\mathbf{S_2S_2}$ interactions.

According to studies in the non-spinning case [17, 24, 25], the TaylorT4 appears to be a very good approximant up to a frequency $\bar{\omega} = \pi\bar{f}_{GW} \simeq 0.1/M$ for the equal mass case, even though its faithfulness seems to worsen for different mass-ratios.

The PN evolution is halted when ω_{orb} reaches the value ω_m that is determined by comparison with numerical waveforms. For $\omega_{orb} > \omega_m$ (ω_{orb} is monotonic here) the angular frequency is evolved according to

$$\omega_{orb}(t) = \frac{\omega_1}{1 - t/T_A} + \omega_0, \qquad \omega_m < \omega_{orb} \text{ and } t < t_{rac}\,, \qquad (4)$$

where the three unknown parameters $\omega_{0,1}$ and T_A are fixed by requiring that $\omega_{orb}, \dot{\omega}_{orb}$ and $\ddot{\omega}_{orb}$ be continuous at the matching point defined by ω_m, and t_{rac} will be defined shortly. In this phenomenological phase the amplitude of the waveform is evolved according to the lowest order formula eq.(3), with ι and α frozen to their values at the instant of time when $\omega_{orb} = \omega_m$ (and keeping as usual $v = (M\omega_{orb})^{1/3}$) [1].

After this phenomenological phase, in order to smoothen the evolution of ω_{orb}, which should settle at the ring-down value ω_{rd}, it turned out useful to join the time evolution of eq.(4) with a smoothing function

$$\omega_{orb}(t) = \omega_{rd} - \left(1 - \frac{t}{T_A}\right)^2 \omega_2 \qquad t_{rac} < t \text{ and } \omega_{orb} < \omega_{mrd} \equiv 0.8\,\omega_{rd} \qquad (5)$$

[1] Clearly the approximation of constant ι and α in the phenomenological phase is a brutal one and it can be badly wrong in the case of strongly precessing systems (like systems with $\nu \ll 1$ where the more massive body has (nearly-)extremal spin not-aligned with the orbital angular momentum. However in those cases a phenomenological formula in the same spirit of eq. (4) can be written for ι and α.

where t_{rac}, ω_2 are fixed by requiring continuity of ω_{orb} and $\dot{\omega}_{orb}$. For $\omega_{orb} > \omega_{mrd} = 0.8\,\omega_{rd}$ the waveforms is described by the damped exponential

$$h_{2,2}^{(rd)} = e^{-t/\tau}\left[A\cos(\omega_{rd}t) + B\sin(\omega_{rd}t)\right] , \tag{6}$$

where A and B are constant to be determined requiring the continuity of $h_{2,2}$. For each multipolar mode defined by a pair l, m there is an infinity of overtones with increasing damping factor, but for our practical purposes retaining only the first overtone is enough.
Note that the degree of continuity of the waveform across the phenomenological -ring-down phase transition is not affected by changing the value of ω_{mrd}, but only by the number of overtones admitted to describe the ring-down phase in eq. (6), as it happens in the EOBNR construction [26].

The values of the ring-down frequencies and damping factors of the three lowest overtones of the $l \leq 4$ modes can be read from [27] as a function of the mass and spin of the final object created by the merger of the binary system. The final mass is determined by the algebraic sum of the constituents' masses and the negative binding energy once ω_m is reached, and the final spin according to the phenomenological formula given in [28].
The damped exponential is attached when ω_{orb} reaches 80% of the ω_{rd} value as it turned out that this specific value allowed the best overlap between analytical and numerical waveform, irrespectively of the initial spins.

The analytical waveforms just described has been quantitatively confronted with the numerically generated ones by computing the overlap integral

$$I_{h_1,h_2} \equiv \int h_1(f)h_2^*(f)\,df , \tag{7}$$

maximized over initial phase and time of arrival. The angular frequency ω_m has been determined by a first set of numerically produced short waveforms (4-6 cycles long) by picking the value which allowed a maximum overlap integral with the phenomenological waveform, with a determination precision of $\pm 10^{-4}/M$.

The numerical waveforms used in the present work have been generated with `MayaKranc`. The grid structure for each run consisted of 10 levels of refinement provided by `CARPET` [29], a mesh refinement package for `CACTUS` [30]. Sixth-order spatial finite differencing was used with the BSSN equations implemented with `Kranc` [31]. The outer boundaries are located at $317M$ and the finest resolution is $M/77$. Waveforms were extracted at $75M$. A few waveforms were generated at resolutions of $\{M/64, M/77, M/90\}$, and convergence consistent with our fourth order code is found. The short (long) runs showed a phase error on the order of $5 \cdot 10^{-3}$ ($5 \cdot 10^{-2}$) radians and an amplitude error of $\approx 2\%$ ($\approx 5\%$). We expect similar accuracy in all runs performed.

The numerical waveforms describe the $l = 2$, $m = 2$ mode and consist of two sets, both with equal mass and spin magnitudes $|\mathbf{S_1}| = |\mathbf{S_2}| = 0.6m_1^2$. The first set consisted in 24 few-cycle-long waveforms, representing mostly the merger and ring-down phases of a coalescence. They have been used to fix the values of ω_m for the corresponding values of initial spins. One of the binary constituent had initial spin $\mathbf{S}_2/m_1^2 = (-0.6, 0, 0)$ in the reference frame in which the initial $\hat{\mathbf{L}} = (0, 0, 1)$. The different values of the first dimension-less spin have been obtained by rotating the $(0, 0, 0.6)$ vector by 15 degrees in the x-z plane. Once determined the values of ω_m for each waveform, the value of the matching frequency ω_m for generic spins has been determined by assuming an analytical dependence of ω_m on $\mathbf{S}_{1,2}$, according to

$$\begin{aligned} M\omega_m = {} & a_0 + a_1(S_{1z} + S_{2z}) + a_2\delta(S_{1z} - S_{2z}) + a_3(S_1S_2)_\perp + a_4(S_{1\perp}^2 + S_{2\perp}^2) + \\ & a_5\delta(S_{1\perp}^2 - S_{2\perp}^2) + a_6(S_{1z}^2 + S_{2z}^2) + a_7(S_{1z}S_{2z}) + a_8\delta(S_{1z}^2 - S_{2z}^2) + \dots , \end{aligned} \tag{8}$$

Journal of Physics: Conference Series **243** (2010) 012007 doi:10.1088/1742-6596/243/1/012007

where the suffix $\perp$ stands for projection onto the plane perpendicular to $\hat{\mathbf{L}}$ and higher power of the spin variables have been neglected. The spin components are understood in a frame where the orbital angular momentum is along the z-axis and their values are time-dependent, thus inducing a mild time dependence in ω_m. Note that because of the dependence of the ω_{orb} evolution equation on $\mathbf{L}$, the spin components parallel to the orbital angular momentum enter already at linear level, whereas the dependence on the spin components in the plane of the orbit starts from the quadratic level. The a_i coefficients may depend on the symmetric mass ratio ν, but it is assumed here that they can be analytically expanded around their value at $\delta \equiv \sqrt{1-4\nu} = (m_1 - m_2)/(m_1 + m_2) = 0$, according to

$$a_i(\delta) = a_i + \delta\, a_i^{(1)} + \delta^2 a_i^{(2)} + \ldots . \quad (9)$$

Since the set of simulations analyzed in this work do not include the case of unequal masses, it has been possible to fit only the first term in the expansion (9) for all the coefficients. [2]

Given the specifics of the simulations we used (all having $\delta = 0$ and $\mathbf{S_2} \cdot \hat{\mathbf{L}} = 0$ initially), it has been possible to determine only the coefficients a_0, a_1, a_3, a_4, a_6. These values have then been tested against a second set of numerical waveforms, consisting of 8 long waveforms (12-15 cycles long), where all the three phases (inspiral, merger and ring-down) play a role. In particular in the long waveform case, even with the initial condition $\mathbf{S}_2 \cdot \hat{\mathbf{L}} = 0$, the dynamical evolution has lead to generic $\mathbf{S}_2$ at the time of the onset of the phenomenological phase. Thus the determination of the unknown coefficients $a_{2,7,8,...}$ will be necessary for an accurate estimation of ω_m. Additional numerical simulations are required to obtain these extra information and thus improve our model.
The result of the comparison between analytical and numerical waveform is the subject of the next section.

3. Results

The analytical waveforms have been calibrated by comparing them with 24 short numerical simulations, with the result qualitatively shown in figs. 1,2 and quantitatively reported in tab. 1.

The determination of the ω_m's giving the best overlap for different spin values allowed to evaluate some of the coefficients in the phenomenological formula (8), as given in tab. 2.
The coefficients $a_{2,5,7,8}$ cannot be determined by the analysis performed because the terms they multiply vanish identically for the 24 simulations considered. The a_0 coefficient has been determined by comparison with a non-spinning waveform[3].

Having fixed the value of ω_m with some generality, it is possible to generate analytical waveforms with any specific initial condition without tuning any single parameter: the value of ω_m will be determined analytically via eq. (8) with the unknown coefficients arbitrarily set to zero. It is then possible to generate waveforms with no tunable parameters for comparison with the second set of long, numerically generated waveforms, even though a wider range of initial conditions would be needed for a more accurate determination of ω_m in the most generic case. Such comparison is summarized qualitatively in fig. 3 and quantitatively in tab. 3.

[2] Note that from the explicit formula for ω_{orb}, see e.g. [21], symmetric combinations of the spin components can only depend analytically on the symmetric mass ratio ν, and thus will depend on even power of δ, whereas anti-symmetric combinations do not appear for $\delta = 0$.

[3] Note that given the moderate value of the spin magnitude, the plane of the orbit relative to these waveforms are only "mildly" precessing.

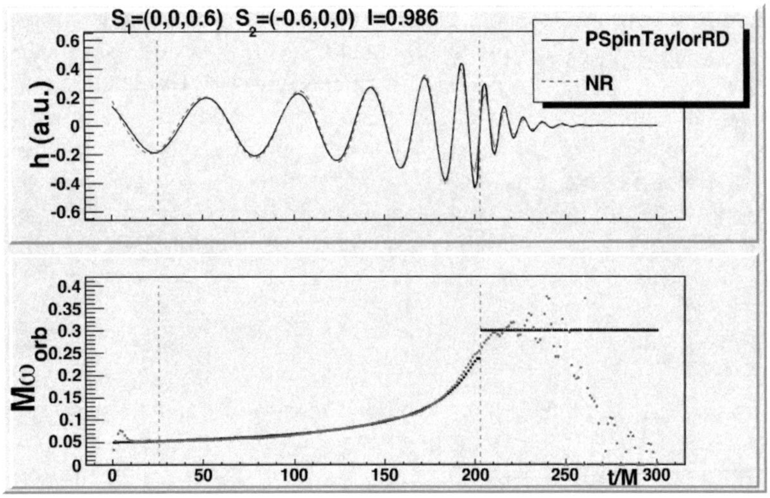

Figure 1. Comparison between the $l = 2, m = 2$ mode of analytical (black, solid) and numerical (red, dashed) equal mass simulations: time-domain waveforms are shown in the upper frame, the lower frame shows ω_{orb} (given respectively by its PN value, by eq.(4) and by ω_{rd} in the three different phases) vs. the instantaneous frequency of the numerical waveform (computed as $\mathrm{Im}(\dot{h}_+ + i\dot{h}_\times)/|h_+ + ih_\times|$, being h_+ ($h_\times$) the plus (cross) polarization of the numerical waveform). Initial spin configuration: $\mathbf{S_1}/m_1^2 = (0, 0, 0.6)$, $\mathbf{S_2}/m_2^2 = (-0.6, 0, 0)$. Vertical dashed lines mark the onset of the phenomenological phase parametrized by eq. (4) and the ring down phase parametrized by eq. (6). The resulting overlap integral is 0.986.

4. Conclusions

We presented an analytical method to produce complete gravitational waveforms from spinning coalescing binaries. The free parameters of the model are the values of the orbital frequency at the transition from the inspiral to the phenomenological phase and from the phenomenological to the ring-down phase. They have undergone a calibration process by confronting the analytical waveforms with numerically generated ones and in particular the accurate determination of the first of such parameters (ω_m) turned to be crucial for a satisfactory waveform construction.
For the calibration process a first set of (short) waveforms has been employed, obtaining overlap factors ranging between 0.97 and 0.99. Once the calibration has been obtained, the analytical waveforms, with no parameter to tune, had been confronted with a second set of (long) numerical waveforms and overlap factors ranging from 0.95 to 0.99 have been obtained. Better overlap is expected once a larger number of available numerical waveforms makes it possible to have a more solid estimation of the matching frequency ω_m for generic spins.
The method illustrated here can be extended to produce other modes than the $l = 2, m = 2$ for which it has been tested. The $m = \pm 2$ quadrupolar modes are the only non-vanishing modes for head-on observation of a coalescing system, whereas to build physical waveforms for generic angle between the source and the observer all other modes are required, at least with $l \leq 6$. In particular, as the quality of the numerical waveforms describing such modes improves, it would be interesting to verify if the calibration performed here on the dominant quadrupolar mode allow a good description of other modes.
The programs generating these phenomenological waveforms are written in the C language and are available from the LIGO Analysis Library (LAL) [32].

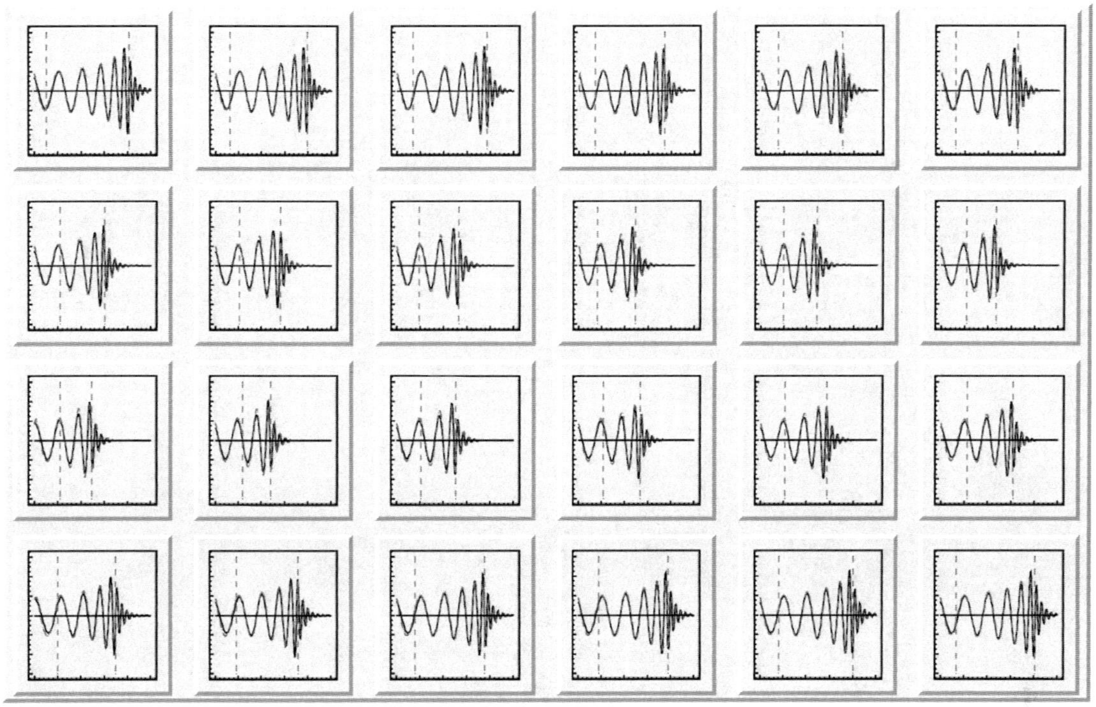

Figure 2. Summary of the comparison between the $l = 2, m = 2$ mode of analytical (black, solid) and numerical (red, dashed) waveforms. Initial spin configuration: $\mathbf{S_2}/m_2^2 = (-0.6, 0, 0)$, $\mathbf{S_1}$ lies in the x-z plane. A rotation of $\mathbf{S_1}$ by 15 degrees in the x-z plane takes from one plot to the following one, the first plot is the same as in fig. 1. Vertical dashed lines, x- and y-axis are the same as in fig. 1.

Acknowledgments
It is a pleasure to thank the organizers of the GWDAW-14 conference in Rome for the stimulating scientific environment created by the meeting, as well as Evan Ochsner for useful comments on the paper. R.S. wishes to thank the Physics Department of the University of Massachusetts at Amherst for kind hospitality during the preparation of part of this work. S.F. and L.C. wishes to thank the Università di Urbino for kind hospitality and the Istituto Nazionale di Fisica Nucleare (INFN) for support during the early stage of this work. This work was supported by the INFN and by the NSF grants PHY-0653550, PHY-0925345, PHY-0941417 and PHY-0903973.

References

[1] L. Blanchet, Living Rev. Rel. **9** (2006) 4.
[2] S. A. Teukolsky, Astrophys. J. **185** (1973) 635.
[3] F. Pretorius, Phys. Rev. Lett. **95** (2005) 121101 [arXiv: gr-qc/0507014].
[4] M. Campanelli, C.O. Lousto, P. Marronetti, and Y. Zlochower, Phys. Rev. Lett. **96** (2006) 111101 [arXiv:gr-qc/0511048].
[5] J.G. Baker, J. Centrella, D. Choi, M. Koppitz and J. van Meter, Phys. Rev. Lett. **96** (2006) 111102 [arXiv:gr-qc/0511103].
[6] F. Herrmann, I. Hinder, D.M. Shoemaker, P. Laguna and R.A. Matzner, Phys. Rev. D **76** (2007) 084032 [[arXiv:0706.2541]].

Journal of Physics: Conference Series **243** (2010) 012007 doi:10.1088/1742-6596/243/1/012007

$acos(S_1\hat{L})[^o]$	Overlap	$M\omega_m \times 10^2$	$M\omega_{rd} \times 10^2$
0	0.986	5.29	30.0
15	0.991	5.37	29.5
30	0.991	5.44	28.8
45	0.986	5.49	29.0
60	0.977	5.55	28.0
75	0.975	5.58	28.0
90	0.983	6.07	26.9
105	0.980	6.13	26.5
120	0.981	6.18	26.1
135	0.987	6.00	24.7
150	0.983	6.36	26.4
165	0.984	6.59	26.0
180	0.985	6.47	26.3
195	0.980	6.70	27.0
210	0.980	6.34	27.5
225	0.979	6.29	27.5
240	0.978	6.08	29.9
255	0.977	6.03	28.9
270	0.971	5.68	29.4
285	0.977	5.62	29.6
300	0.980	5.48	30.4
315	0.985	5.51	30.3
330	0.988	5.44	30.3
345	0.988	5.36	29.8

Table 1. Values of the overlap integral between the analytical and the 24 short numerical waveforms. For reference, the values of ω_m maximizing the overlap and of ω_{rd} are reported.

Coeff.	$\times 10^{-2}$
a_0	5.480
a_1	-0.97
a_3	0.083
a_4	0.47
a_6	0.80

Table 2. Coefficients of eq.(8) as determined by comparison of the analytical waveforms with 24 short numerical waveforms.

[7] W. Tichy and P. Marronetti, Phys. Rev. D **78** (2008) 081501 [arXiv:0807.2985 [gr-qc]].
[8] M.A. Scheel, M. Boyle, T. Chu, L.E. Kidder, K.D. Matthews, H.P. Pfeiffer Phys. Rev. D **79** (2009) 024003 [arXiv:0810.1767 [gr-qc]].
[9] J. G. Baker, W. D. Boggs, J. Centrella, B. J. Kelly, S.T. McWilliams and J.R. van Meter, Phys. Rev. D **78** (2008) 044046 [arXiv:0805.1428 [gr-qc]].
[10] P. Ajith *et al.*, arXiv:0909.2867 [gr-qc].
[11] C.O. Lousto, H. Nakano, Y. Zlochower and M. Campanelli, Phys. Rev. D **81** (2010) 084023 [arXiv:0910.3197 [gr-qc]].
[12] M. Hannam, [arXiv:0901.2931 [gr-qc]].
[13] S. Husa, Eur. Phys. J. ST **152** (2007) 183–207.
[14] F. Pretorius, "Physics of Relativistic Objects in Compact Binaries: From Birth to Coalescence", Springer, Heidelberg (Germany), 2009 [arXiv:0710.1338].

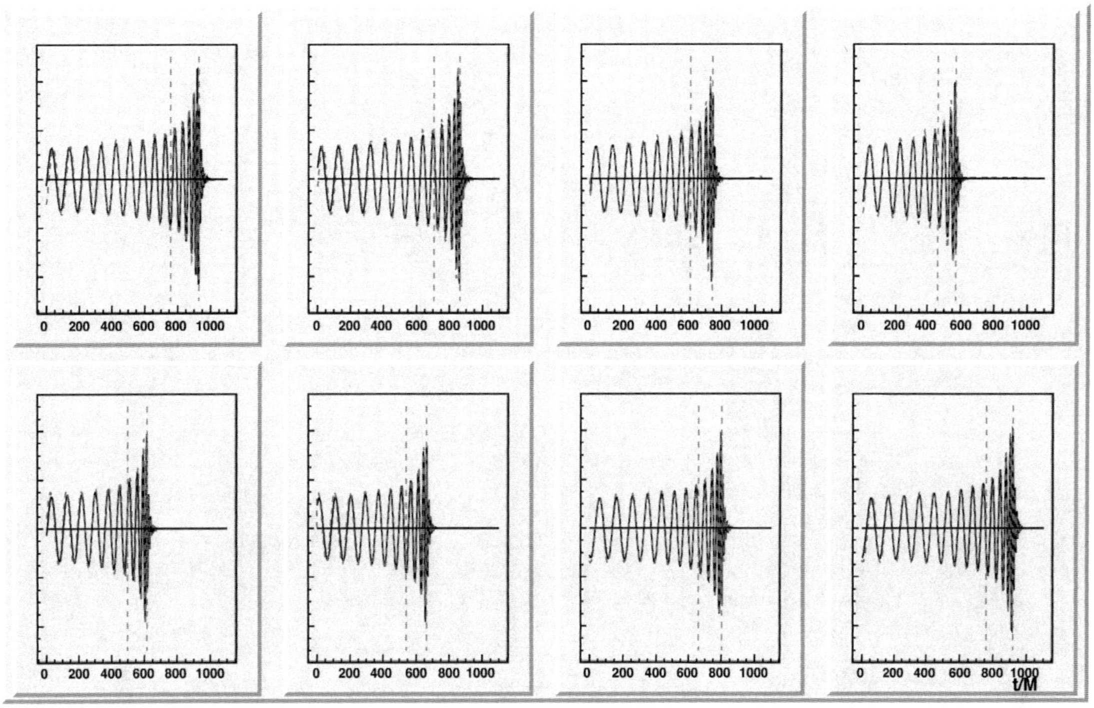

Figure 3. Comparison between the $l = 2, m = 2$ mode of analytical and long numerical waveforms ($m_1 = m_2$). Initial spin configuration: $\mathbf{S_2}/m_2^2 = (-0.6, 0, 0)$, $\mathbf{S_1}$ lies in the x-z plane. A rotation of $\mathbf{S_1}$ by 45 degrees in the x-z plane moves the initial condition of one plot to the following one, the first plot refers to the initial condition $\mathbf{S_1}/m_1^2 = (0, 0, 0.6)$. Vertical dashed lines mark the onset of the phenomenological phase and the ring-down phase, x-axis is time in units of M.

$acos(S_1\hat{L})[^o]$	Overlap	$M\omega_m \times 10^2$	$M\omega_{rd} \times 10^2$
0	0.9508	5.14	29.2
45	0.9761	5.30	28.6
90	0.9935	5.50	27.2
135	0.9882	5.72	24.9
180	0.9499	5.71	25.4
225	0.9605	5.70	25.9
270	0.9934	5.39	26.7
315	0.9698	5.23	28.3

Table 3. Values of the overlap integral between the analytical and the 8 long numerical waveforms. For comparison with the "short" case the values of ω_m and ω_{rd} are reported.

[15] LIGO Scientific Collaboration and Virgo Collaboration, "Search for gravitational-wave inspiral signals associated with short Gamma-Ray Bursts during LIGO's fifth and Virgo's first science run", arXiv:1001.0165.
[16] A. Buonanno and T. Damour, Phys. Rev. D **59** (1999) 084006 [arXiv:gr-qc/9811091].
[17] A. Buonanno, G. B. Cook and F. Pretorius, Phys. Rev. D **75** (2007) 124018 [arXiv:gr-qc/0610122].
[18] T. Damour and A. Nagar, Phys. Rev. D **79** (2009) 081503 [arXiv:0902.0136 [gr-qc]].
[19] Y. Pan, A. Buonanno, L. T. Buchman, T. Chu, L. E. Kidder, H. P. Pfeiffer and M. A. Scheel, arXiv:0912.3466 [gr-qc].
[20] P. Ajith *et al.*, Class. Quant. Grav. **24** (2007) S689 [arXiv:0704.3764 [gr-qc]].
[21] A. Buonanno, B. Iyer, E. Ochsner, Y. Pan and B. S. Sathyaprakash, Phys. Rev. D **80** (2009) 084043 [arXiv:0907.0700 [gr-qc]].
[22] E. Racine, A. Buonanno and L. E. Kidder, Phys. Rev. D **80** (2009) 044010 [arXiv:0812.4413 [gr-qc]].
[23] K. G. Arun, A. Buonanno, G. Faye and E. Ochsner, Phys. Rev. D **79** (2009) 104023 [arXiv:0810.5336 [gr-qc]].
[24] J. G. Baker, J. R. van Meter, S. T. McWilliams, J. Centrella and B. J. Kelly, Phys. Rev. Lett. **99** (2007) 181101 [arXiv:gr-qc/0612024].
[25] M. Boyle *et al.*, Phys. Rev. D **76** (2007) 124038 [arXiv:0710.0158 [gr-qc]].
[26] A. Buonanno, Y. Pan, J. G. Baker, J. Centrella, B. J. Kelly, S. T. McWilliams and J. R. van Meter, Phys. Rev. D **76** (2007) 104049 [arXiv:0706.3732 [gr-qc]].
[27] E. Berti, V. Cardoso and C. M. Will, Phys. Rev. D **73** (2006) 064030 [arXiv:gr-qc/0512160].
[28] E. Barausse and L. Rezzolla, Astrophys. J. **704** (2009) L40 [arXiv:0904.2577 [gr-qc]].
[29] E. Schnetter, S.H. Hawley and I. Hawke, Class. Quant. Grav (2004) **21** 1465–1488 [arXiv:0310042 [gr-qc]].
[30] Cactus Computational Toolkit home page: `http://www.cactuscode.org`.
[31] S. Husa, I. Hinder and C. Lechner, Computer Physics Communications **174** (2006) 983-1004 [arXiv:0404023 [gr-qc]].
[32] https://www.lsc-group.phys.uwm.edu/daswg/projects/lal.html

Journal of Physics: Conference Series **243** (2010) 012008 doi:10.1088/1742-6596/243/1/012008

Compact binary waveform recovery from the cross-correlated data of two detectors by matched filtering with spinning templates

L. Veréb[1,2], Z. Keresztes[1,2⋆], P. Raffai[3†], Zs. Udvari[1], M. Tápai[1,2], L. Á. Gergely[1,2‡]

Abstract. We investigate whether the recovery chances of highly spinning waveforms by matched filtering with randomly chosen spinning waveforms generated with the LAL package are improved by a cross-correlation of the simulated output of the L1 and H1 LIGO detectors. We find that a properly defined correlated overlap improves the mass estimates and enhaces the recovery of spin angles.

[1] Department of Theoretical Physics, University of Szeged, Tisza Lajos krt 84-86, Szeged 6720, Hungary
[2] Department of Experimental Physics, University of Szeged, Dóm tér 9, Szeged 6720, Hungary
[3] Institute of Physics, Loránd Eötvös University, Pázmany Péter s 1/A, Budapest 1117, Hungary

E-mail: *zkeresztes@titan.physx.u-szeged.hu †praffai@bolyai.elte.hu
‡gergely@physx.u-szeged.hu

1. Method

Recovering a gravitational wave pattern in the noisy detector output is a difficult problem. In this work we present our attempt to recover a gravitational spinning waveform h_{inj} immersed in S5 LIGO-like noise by a modified matched filtering method. We calculate the overlap [1] between the noisy injection $h_{n,i}$ and spinning templates $h_{template}$ as

$$O\left[h_{n,i}, h_{template}\right] = \frac{\langle h_{n,i}|h_{template}\rangle}{\sqrt{\langle h_{n,i}|h_{n,i}\rangle \langle h_{template}|h_{template}\rangle}} , \tag{1}$$

with

$$\langle h_1|h_2\rangle = 4Re\int_{f_{\min}}^{f_{\max}} \frac{\widetilde{h}_1(f)\,\widetilde{h}_2^*(f)}{S_n(f)}df , \tag{2}$$

where tilde and star denote Fourier transform and complex conjugate, respectively, and $S_n(f)$ is the power spectral density of the noise. We choose $f_{\min} = 50$Hz and $f_{\max} = 600$Hz, lying in the best sensitivity band of the LIGO detectors, also the post-Newtonian prediction for the waveform is quite accurate there.

On theoretical grounds, infinite long data series would be required for exact determination of the power spectrum. In order to achieve stability of the power spectral density of order of 1%, we would like to have at least 100 periods of the lowest frequency, therefore a minimal length of

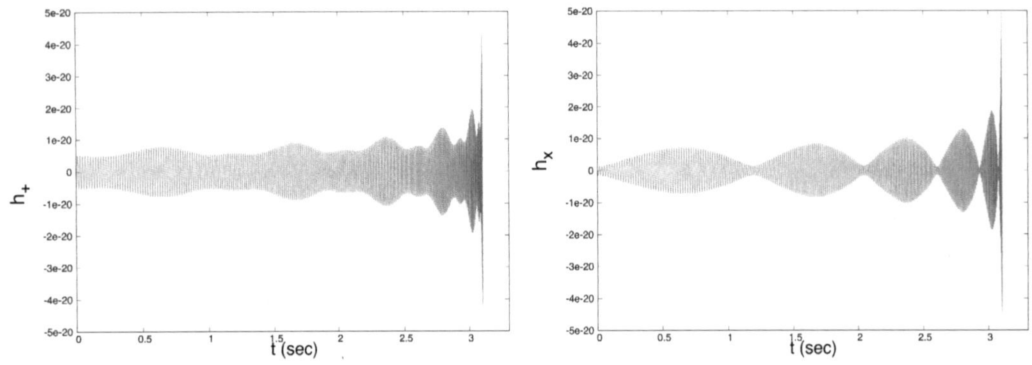

Figure 1. The polarization h_+ (left) and $h_\times$ (right) of the spinning waveforms.

the templates of 2 sec was imposed. We compute the above overlap both for the Hanford and Livingston detectors (O^H and O^L, respectively).

We also defined a *correlated match*, the overlap O^{HL}_{corr}, using the correlations of the Hanford and Livingston signals for $h_{n,i}$, $h_{template}$ and noises. (The correlation of h_1 and h_2 in Fourier space is defined as $\widetilde{H}(f) = \widetilde{h}_1(f)\widetilde{h}_2^*(f)$.)

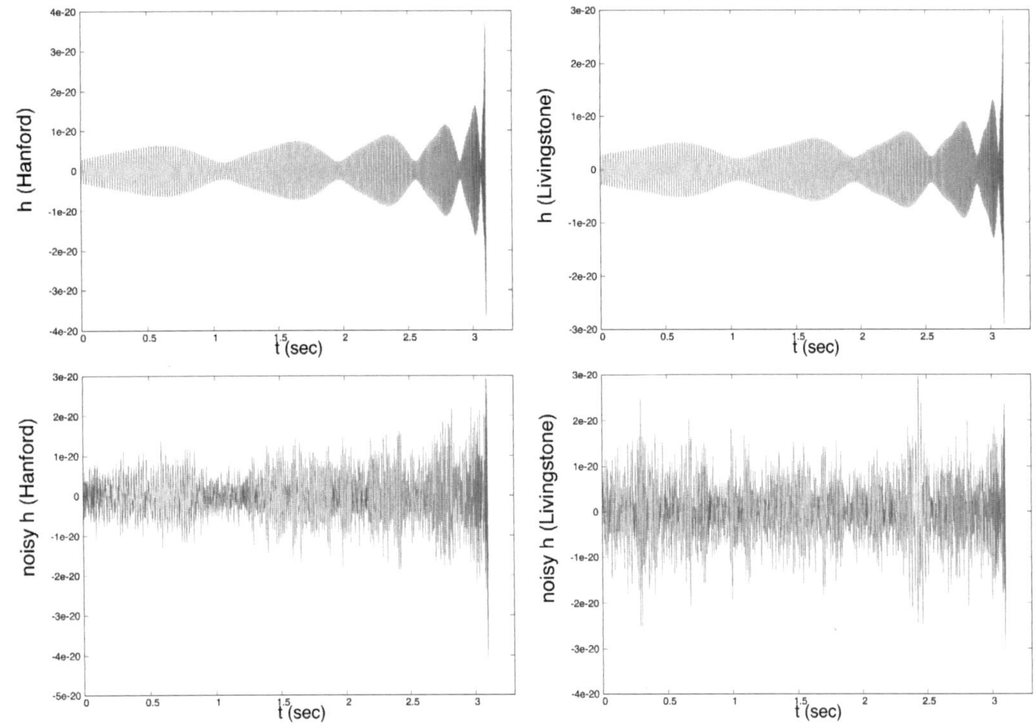

Figure 2. Spinning waveforms at Hanford (top left) and Livingston (top right), mixed in LIGO S5-like noise (bottom).

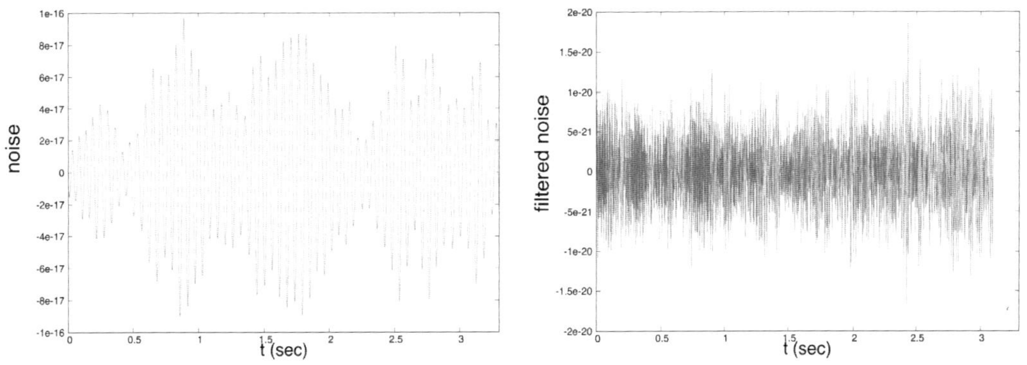

Figure 3. Characteristic unfiltered (left) and filtered (right) LIGO noises.

Table 1. The parameters of the gravitational wave polarizations $(h_+, h_\times)$ of the injected signal. Masses m_i, magnitude of the dimensionless spins χ_i, initial direction of the spin vectors given by $\cos\kappa_i$ and ψ_i in the frame with the line of sight on the z-axis, and the initial orbital angular momentum $\mathbf{L_N}$ in the x-z plane, spanning the angleΘ , and the distance d_L of the source. The initial phase is 0; the initial time being also fixed to 0.

name	$m_1(M_\odot)$	$m_2(M_\odot)$	χ_1	χ_2	$\cos\kappa_1$	ψ_1	$\cos\kappa_2$	ψ_2	Θ	$d_L\,(Mpc)$
injection	3.553	3.358	0.983	0.902	0.984	1.109	0.978	0.957	1.430	1

Table 2. The angles θ, φ and ψ (polarization angle) give the relation between a fictitious Earth-centered detector and the source frame. For the actual detector positions further rotation have to be taken into account [5]. These angles are also necessary for computing the antenna functions $(F_+, F_\times)$. The gravitational signal is $h = h_+F_+ + h_\times F_\times$.

name	φ	θ	ψ
injection	3.657	0.278	0.000

For the two detectors we also employed the quantity O^{HL}, according to Ref. [2]:

$$O^{HL} = \frac{\langle h_{n,i}|h_{template}\rangle_H + \langle h_{n,i}|h_{template}\rangle_L}{\sqrt{\left(\langle h_{n,i}|h_{n,i}\rangle_H + \langle h_{n,i}|h_{n,i}\rangle_L\right)\left(\langle h_{template}|h_{template}\rangle_H + \langle h_{template}|h_{template}\rangle_L\right)}} \,. \tag{3}$$

Therefore we could compare four type of matches.

The spinning waveforms, based on Ref. [3], were computed by using SpinTaylor Ref. [4], while the antenna functions based on Ref. [5] with the XLALComputeDetAMResponse() function in DetResponse.c under the LAL package Ref. [4].

Table 1 contains the parameters of the injected waveform. On Fig 1 the two polarizations of this waveform h_+ and $h_\times$ are shown. The parameters characterizing the source and the detector orientation, which are necessary for computing the antenna functions F_+ and $F_\times$ are given in Table 2. The waveforms appearing at the Hanford and Livingston detectors are plotted on Fig 2 (top line), while the bottom line on Fig 2 shows the signals immersed in S5 LIGO-like noise. Some characteristic unfiltered and filtered noises can be seen on Fig 3.

First we calculated the four type of overlaps between the noisy injection and the injected signal itself, finding $O^H\left[h_{n,i}, h_{injection}\right] = 0.8392$, $O^L\left[h_{n,i}, h_{injection}\right] = 0.9064$, $O^{HL}\left[h_{n,i}, h_{injection}\right] = 0.8662$ and $O^{HL}_{corr}\left[h_{n,i}, h_{injection}\right] = 0.8646$. Next we defined the following auxiliary quantity for all overlaps

$$\sigma = \left| 1 - \frac{O\left[h_{n,i}, h_{template}\right]}{O\left[h_{n,i}, h_{injection}\right]} \right| , \tag{4}$$

vanishing for the injected template. We searched for templates with the various σ-s less then 0.1.

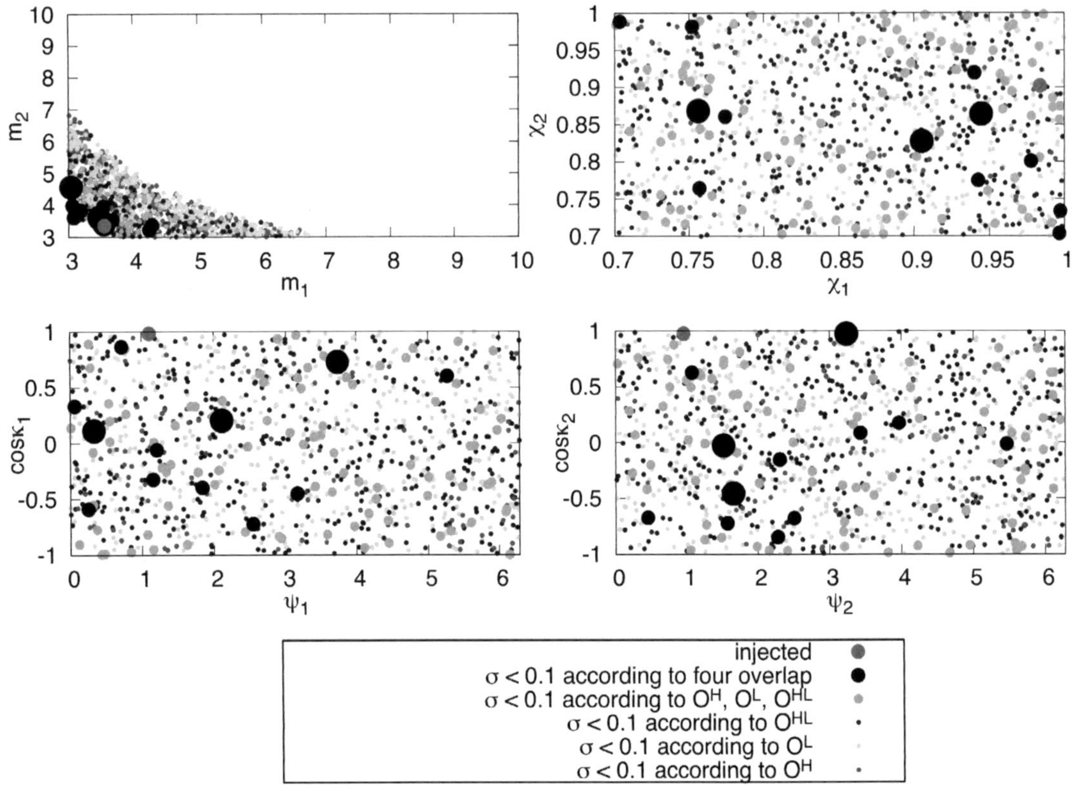

Figure 4. The parameters of the injected signal (red) and the templates having $\sigma < 1$ according to O^H (green), O^L (yellow), O^{HL} (dark blue) and all of these (light blue) and when the correlated overlap was also taken into account (black). The big dots among the black represent those templates that has the smallest values accordint to the correlated overlap.

The templates were chosen with parameters in the ranges: masses $m_i \in 3 \div 10$ M$_\odot$; dimensionless spins $\chi_i \in 0.7 \div 1$; spin angles $\cos\kappa_i$ and ψ_i random; distance d_L, anglesΘ, φ, θ and ψ are fixed identically to the values given in Tables 1 and 2 for the injection, respectively. We also assumed as known the time of signal arrival to the detector. Therefore we varied two mass and six spin parameters altogether.

2. Results

Our analysis is based on matching more than one million templates. The templates with any of the σ^H, σ^L, $\sigma^{HL} < 0.1$ were selected and represented on Fig 4 in the parameter planes (m_1, m_2), (χ_1, χ_2) and $(\cos\kappa_i, \psi_i)$ for both spins. The green, yellow and navy dots (for grayscale see the legend) represent templates with required values of σ^H, $\sigma^L$and σ^{HL}, respectively; templates with all three values of σ below the threshold are plotted with larger turqoise dots. Twelve even larger black dots show templates with all four σ-s (including σ^{HL}_{corr}) below the threshold. The

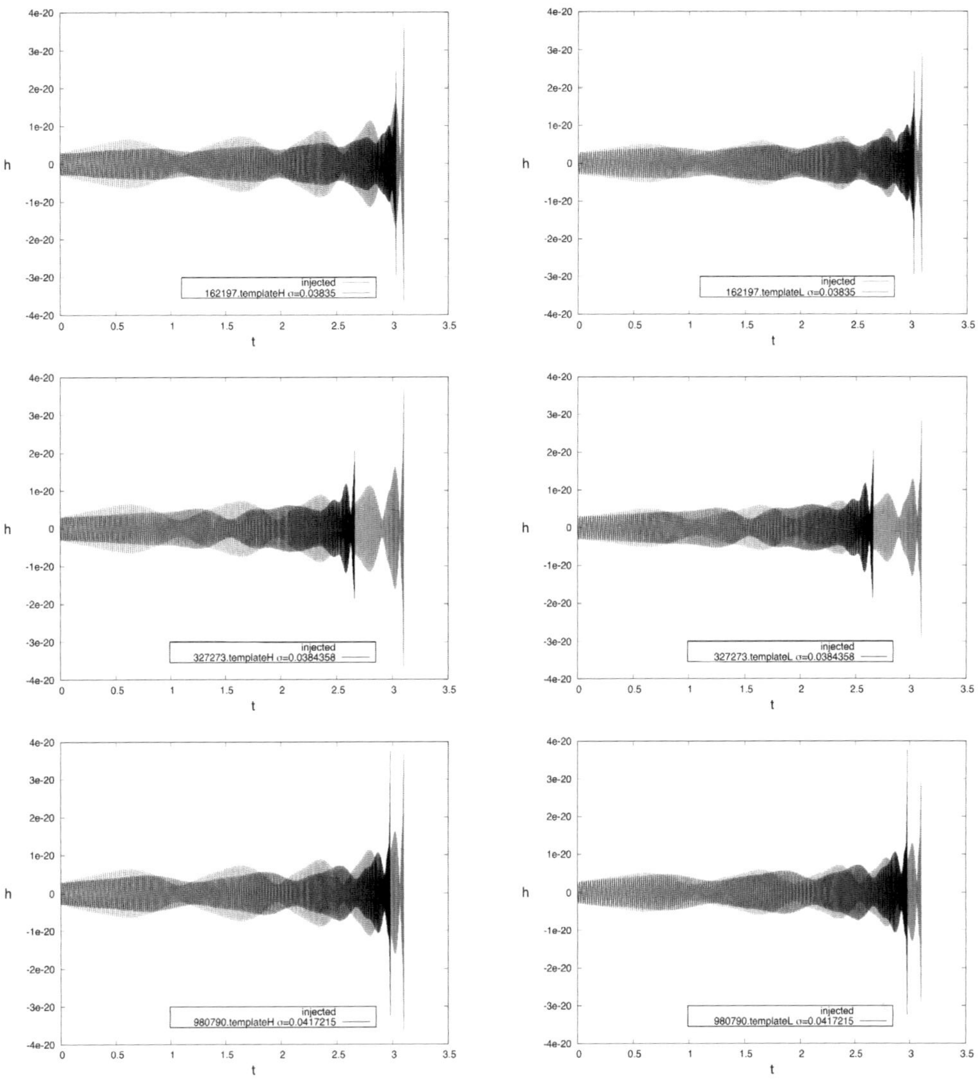

Figure 5. Best three templates (black) as they would appear at Hanford (left) and Livingston (right), compared to the respective injected signals (red).

three largest of them have the lowest value of σ_{corr}^{HL}. The parameters of the best three templates are shown in Table 3, and they are plotted on Fig 5 as they would appear at the Hanford and Livingston detectors.

Discussion: The masses are reasonably well recovered, although sligtly overestimated with any of the σ^H, σ^L, σ^{HL}. The additional monitoring of the correlated match σ_{corr}^{HL} imposes however a selection effect which *improves the estimation of the masses* (top left panel of Fig 4). While the recovery of the spin magnitudes is still problematic (top right panel of Fig 4), the estimation of the spin angles seems slightly improved by the use of of the correlated match σ_{corr}^{HL}. (Black dots exhibit a belt-like structure on both bottom panels of Fig 4.) How relevant is this feature statistically is currently under investigation.

Acknowledgements: This work was supported by the Polányi Program of the Hungarian National Office for Research and Technology (NKTH) and the Hungarian Scientific Research

Table 3. The parameters of the templates represented by the four big black points on Fig 4.

name	$m_1(M_\odot)$	$m_2(M_\odot)$	χ_1	χ_2	$\cos\kappa_1$	ψ_1	$\cos\kappa_2$	ψ_2
162197.template	3.471	3.612	0.905	0.828	0.114	0.330	−0.026	1.515
327273.template	3.038	4.572	0.944	0.864	0.729	3.736	−0.455	1.652
980790.template	3.549	3.426	0.756	0.868	0.206	2.110	0.974	3.237
281270.template	3.057	3.989	0.703	0.989	−0.057	1.209	−0.721	1.569

Fund (OTKA) grant no. 69036.

[1] B. Vaishnav, I. Hinder, F. Herrmann, D. Shoemaker, Phys. Rev. D 76 084020 (2007)
[2] V. Raymond, M. V. van der Sluys, I. Mandel, V. Kalogera, C. Roever, N. Christensen, arXiv:0912.3746 (2009)
[3] A. Buonanno, Y. Chen, M. Vallisneri, Phys. Rev. D 67 104025 (2003); Erratum-ibid. D 74 029904 (2006)
[4] https://www.lsc-group.phys.uwm.edu/daswg/projects/lal.html
[5] W. G. Anderson, P. R. Brady, J. D. E. Creighton, É. É. Flanagan, Phys. Rev. D 63 042003 (2001)

Journal of Physics: Conference Series **243** (2010) 012009 doi:10.1088/1742-6596/243/1/012009

A discrete resampling technique to correct for Doppler effect in continuous gravitational wave search

S.Braccini[a,*], **G.Cella**[a], **I.Ferrante**[a,b], **D.Passuello**[a], **O.Torre**[a,c]
[a]INFN, Sezione di Pisa, Italy; [b]Università di Pisa, Italy; [c]Università di Siena, Italy;
*Corresponding Author (e-mail: stefano.braccini@pi.infn.it)

Abstract. The detection of continuous gravitational waves has to deal with the Doppler effect induced by the Earth motion with respect to the source. This frequency shift must be taken into account to recover the signal energy as a monochromatic peak with a high signal-to-noise ratio. The correction to be applied to the antenna output depends on the source sky direction, on the source spin and also on the spin-down rate. Since they are, in general, unknown parameters a large computational effort is necessary to correct for any possible value. A correction technique independent of the source frequency is discussed here. The method consists to anticipate or delay the antenna clock by removing or doubling one of its digital signal sample, in order to mantain the clock of the moving observer well locked to the rest one. The method, which requires just a little computational effort, appears to be very effective for "semi-targeted" searches, where the source direction is known but the emission frequency is not.

1. Introduction

Rotating neutron stars are expected to emit permanently gravitational waves mainly at the double of their rotational frequency (from fractions of Hz up to several hundreds of Hz). Long observation times are necessary in order to distinguish the peak at the emission frequency in the antenna spectral floor, since the signal-to-noise ratio scales as the square root of the measurement period [1]. The small spectral peak is smeared, and thus buried in the noise, by the phase modulation induced by the Doppler effect due to the antenna motion, and by the phase shift induced by the slowing down of the source rotational frequency *spin-down*. This last effect, whose values range between 10^{-17} and a few 10^{-10} Hz/s, is induced by the energy losses due to emissions of electromagnetic and gravitational waves and can be measured by the electromagnetic observations of the galaxy Pulsars. The correction for recovering a monochromatic peak at the frequency of the source depends on the spin-down rate $\dot{\nu}_0$, on the source sky direction, and on the source emission frequency ν_0. Here, we propose a correction technique that, once fixed the source direction and the spin-down rate, is valid at the same time for all the possible emission frequencies in the detection band. The technique consists in a resampling of the antenna output signal, aimed at synchronizing with a large accuracy the detector time with the source one, so to put the antenna "at rest" with respect to the source. The method, as other techniques [1], is based on a discrete resampling. In particular, in our case, we accelerate (or slow down) the antenna proper time by removing (or doubling) in a timely manner single samples of the detector digitized signal in order to keep the synchronization with

Journal of Physics: Conference Series **243** (2010) 012009 doi:10.1088/1742-6596/243/1/012009

the source clock. The applied correction, and then the possibility to recover a signal peak in the spectrum of the antenna output, is valid for all the possible emission frequencies. If the source direction is known, but the emission frequency ν_0 and the spin-down rate $\dot{\nu}_0$ are unknown (*semi-targeted search*), one can perform several times the resampling, assuming different possible values of the spin-down rates (more precisely of the *fractional spin-down*, $\dot{\nu}_0/\nu_0$), and then search for a peak coming out from the spectral floor in the resampled data streams. In the usual techniques, once fixed the sky direction, the correction must be applied for any possible value of the couple of parameters $\dot{\nu}_0$ and ν_0, with a larger computational effort.

Let us consider a monocromathic signal with amplitude A_0 and frequency ν_0. A Fourier analysis performed on a 1 year-long data bench by an observer at rest with respect to the source exhibits a peak in the spectrum with amplitude A_0 and width $\delta\nu = \frac{1}{1\text{Year}} = 3.17\ 10^{-8}$ Hz. Let us considered the simple case of a detector covering uniformly in one year a circular orbit centered on the Sun (the orientation of the detector is assumed to be constant, that means to neglect any amplitude modulation effect. The detected signal will be:

$$A(t) = A_0 \cos(\omega_0 t + \varphi(t)) \qquad \text{where} \qquad \varphi(t) + \epsilon \sin(\omega_y t) \tag{1}$$

While $\omega_0 = 2\pi\nu_0$, $\omega_y = 2\pi\nu_y$ and ν_y denotes the frequency corresponding to 1 year orbital period. ϵ is the amplitude of the phase variation due to the Doppler effect in one year (modulation index). This phase modulation induces a spread of the spectral line whose half bandwidth, for a source located on the ecliptic plane, is given by:

$$\Delta\nu = \nu_0 \frac{v}{c} \simeq 10^{-4}\nu_0 \tag{2}$$

where v is the orbital speed (around 30 km/s for the Earth), while c is the light speed. Since the energy detected by the antenna (and thus the power spectrum integral) is not modified by the Doppler effect, the square of the signal amplitude will be reduced by a factor of the order of $\frac{\delta\nu}{2\Delta\nu}$. For a wave at a frequency $\nu_0 = 100$ Hz, this means:

$$\frac{A_1^2}{A_0^2} = \frac{\delta\nu}{2\Delta\nu} \simeq \frac{3.17\ 10^{-8}}{2\ 10^{-4}\ 10^2} = 1.58\ 10^{-6} \tag{3}$$

and thus a ratio of the signal amplitudes around $1.25\ 10^{-3}$, with a reduction of signal to noise ratio around 58 dB.

The Doppler effect can be corrected for a given sky direction by resampling the detected signal with an average frequency equal to the source one, but phase modulated so to compensate for phase modulation due to observer motion. The goal is to recover a large fraction of the signal energy in the original FFT bin, so to have a signal to noise ratio similar to the one measured by the rest observer. The accuracy we need in the Doppler compensation can be computed by developing the phase-modulated signal in Bessel functions. It is easy to demonstrate that the term at the source frequency (J_0) keeps the 60% of the signal energy, if the residual phase modulation amplitude, after the Doppler correction, is less than one radian ($\epsilon < 1$ rad). The signal to noise ratio reduction, expressed in amplitude, will be small, less than 6 dB. This is taken as our specification, even if the entire argument can be easily scaled to ask for a better accuracy in signal to noise ratio recovering, without remarkable impact on the computational effort.

2. Correction for the Doppler effect

The method discussed here aims at synchronizing the moving observer clock to the rest one with an accuracy better than a sampling interval $\Delta t = 1/\nu_s$, where ν_s is the sampling frequency of the antenna digitized output signal (typically between 10 and 20 kHz). Let us consider for the

Journal of Physics: Conference Series **243** (2010) 012009 doi:10.1088/1742-6596/243/1/012009

purpose a "virtual" (not physical) sinusoidal wave with a frequency equal to the sampling one, coming from the direction of the source. Its equiphase surfaces are planes perpendicular to the wave vector $\overrightarrow{k}_s$, traveling at the speed of light. In the rest frame, the travelling equiphase plane equation (for a given phase ϕ) is given by:

$$\overrightarrow{k}_s \cdot \overrightarrow{r} - \omega_s t = \phi \tag{4}$$

where $\omega_s = 2\pi\ \nu_s$ and $|\overrightarrow{k}_s| = \omega_s/c$.
Let us consider the family of planes whose phase ϕ is an integer multiple of 2π. Looking at two successive planes in Eq.(4), it straightforward to understand that these planes travel parallel each other, separated by a time Δt (that means by a distance $c\Delta t$), corresponding to a phase $\omega_s \Delta t$. The rest observer that monitors (in a stroboscopic way) at the sampling frequency ν_s, (i.e. each Δt seconds) the positions of this family of planes, sees at any sample time the j_{th} plane takes the place that the $(j+1)_{th}$ one had at the previous sampling time, and to be replaced by the $(j-1)_{th}$ one. In other words, the equiphase planes of these virtual signal at the sampling frequency, appear to shift each other, but the entire family will have at any sample the same positions. We will use these fixed set of planes as a reference grid in the rest frame, where the motion of the moving observer (i.e. of the antenna) is described by a trajectory $\overrightarrow{r}(t)$. Without any loss of generality, one can assume that at time $t = 0$ (start of the analysis) the two clocks (moving and rest one) are synchronized to $\phi = 0$ in the origin of the rest frame. The origin can be chosen coincident with the time-zero position of the antenna trajectory. At a time t, the phase measured by the moving observer is $\overrightarrow{k}_s \cdot \overrightarrow{r}(t) - \omega_s t$ (Eq.(6)). This must be compared with the one detected by the observer at rest in the origin, $\omega_s t$. The dephasing $(\overrightarrow{k}_s \cdot \overrightarrow{r}(t))$ is thus ruled by the moving observer position with respect to the start grid plane where synchronization occurred.
As mentioned above, the goal is to lock the antenna clock to the rest one so to have a time difference not larger than the sampling interval (Δt). When the moving observer crosses one of the two planes nearest to the origin (the first neighbor toward the wave vector or the one on the opposite side), a dephasing $\omega_s \Delta t$ has been cumulated. We must compensate for this by slowing down (or accelerating, depending on the dephase sign - see below) the moving observer proper time, by a time Δt. This compensation is performed in an easy way, just repeating (or deleting) one of the digitized signal sample. In particular, if the observer motion versus is opposite to the wave one (i.e. $\overrightarrow{v} \cdot \hat{n} < 0$) a negative dephasing with respect to the rest clock occurs. This means that the moving clock is anticipating the rest one (since detects in advance the running equiphase plane). The antenna clock has thus to be slowed by repeating a sample of its output signal stream. Vice versa, when $\overrightarrow{v} \cdot \hat{n} > 0$, a delay occurs, and we compensate by removing a sample. The correction just described must be performed each time the moving observer crosses one of the grid planes. In this way the two clocks are always synchronized with an accuracy less or equal of a sampling interval Δt.
The entire argument has been developed considering a virtual wave, coming from the source direction, having a non-physical emission frequency, equal to the digital signal sampling rate (several kHz). For a generic wave with frequency ν_0, the achieved time synchronization between the two clocks, corresponds to a phase locking accuracy of $\omega_0 \Delta t$, i.e. $2\pi\nu_0/\nu_s$. To meet the required 1 rad phase accuracy it thus necessary to operate with a sampling frequency at least 2π times larger than the source one. In gravitational wave antennas, where use of sampling frequencies between 10 and 20 kHz is made, the technique can thus be applied up to several kHz, enough for continuous gravitational waves by semi-targeted analysis, that usually stop around 200 Hz. This leaves a margin for the achievement of a better recovering accuracy and/or a preliminary sampling frequency reduction of the original data stream. In the usual cases the moving observer velocity does not change too much during the crossing between two grid planes. The time to cross two successive equiphase planes ($t_{crossing}$) is well approximated by

their distance ($c\Delta t$), divided by the amplitude of the antenna velocity ($\overrightarrow{v}$) projection along the wave vector ($|\overrightarrow{v} \cdot \hat{n}|$):

$$t_{crossing} = 1/(|\overrightarrow{\beta} \cdot \hat{n}| \; \nu_s) \tag{5}$$

The minimum possible value for $t_{crossing} = 1/(\beta\nu_s)$, occurs when the velocity direction is parallel to the wave vector. This means that the time between the two successive planes (and thus the correction), independently of the sampling frequency, cannot occur before than $1/\beta$ samples (in the case of Earth, where $\beta \simeq 10^{-4}$, one each about ten thousands). Vice versa, when the velocity of the orbit is almost perpendicular to the wave vector, long times between two successive crossings take place.

3. Spin-down correction

The spin-down effect can be described by the Taylor expansion of the rotational frequency of the Neutron star as a function of time:

$$\nu(t) = \nu_0 + \dot{\nu}_0 t + \frac{1}{2}\ddot{\nu}_0 t^2 + ... \tag{6}$$

where $\dot{\nu}_0, \ddot{\nu}_0, ...$ are named *spin-down parameters.* The cumulated phase of the signal, as measured by the observed at rest with respect to the source, is obtained by integrating $\nu(t)$ over time:

$$\varphi(t) = 2\pi \int_0^t \nu(\tau)d\tau = 2\pi \left(\nu_0 t + \frac{1}{2}\dot{\nu}_0 t^2 + \frac{1}{6}\ddot{\nu}_0 t^3 + ...\right). \tag{7}$$

Taking the first order of the expansion, the signal emitted by the Pulsar is

$$S(t) = h_0 \sin\left(\omega_0 t + \frac{1}{2}\dot{\omega}_0 t^2\right) \tag{8}$$

where h_0 is the gravitational wave amplitude, $\omega_0 = 2\pi\nu_0$ and $\dot{\omega}_0 = 2\pi\dot{\nu}_0$, with typical values of $|\dot{\nu}_0|$, ranging between a few 10^{-17} and a few 10^{-10} Hz/s (see [2]). Since the spin-down induces a decrease of the frequency in time, it is clear that in all cases $\dot{\nu}_0 < 0$.

As shown in the following, a downsampling of the data will be done in order to perform an FFT of a few months long data series with an acceptable number of points. In this frame an effective way to correct also for the spin-down (and synchronize the clocks) is to compute at each downsampling time the value of time t' at which the phase detected by the moving observed $\varphi'(t')$ (left term of next equation) is equal to the one of a perfect monochromatic signal, at the same (generic) emission frequency ω_0. This means to impose:

$$\omega_0 \left(\frac{\vec{r}(t') \cdot \hat{n}}{c} + t'\right) + \frac{1}{2}\dot{\omega}_0 \left(\frac{\vec{r}(t') \cdot \hat{n}}{c} + t'\right)^2 = \omega_0 k t_0 \qquad k = 1, \ldots T/t_0 \tag{9}$$

with the usual notations, and where T is the whole observation time. The Doppler effect in the left term is evident. $t' - kt_0$ represents the time difference between the antenna clock and the pure monochromatic signal. Solving Eq.(9), this time difference (δt) can be obtained:

$$\delta t = t' - kt_0 = -\frac{\vec{r}(t') \cdot \hat{n}}{c} + \frac{\varepsilon (kt_0)^2}{1 - \varepsilon k t_0 + \sqrt{1 - 2\varepsilon k t_0}} \tag{10}$$

where $\varepsilon = \dot{\nu}_0/\nu_0$ is the fractional spin-down rate. δt is the delay/anticipation to be compensated, in order to equalize the two phases at the given downsampling time. Obviously, the time shift can be expressed in terms of the number of samples to be shifted in downsampling extraction,

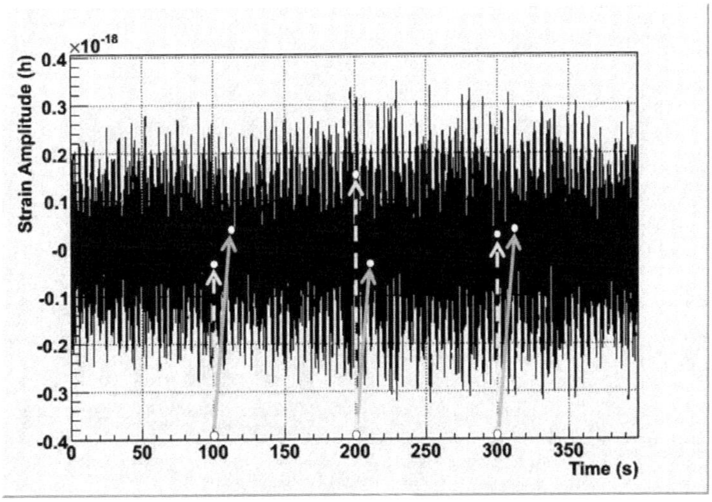

Figure 1. The resampling method: the downsampled stream is generated by taking from the high-frequency stream the sample shifted with respect to the one selected without correction (see dotted arrows), that means exactly at each downsampling time (i.e. 100 s in this case). The number of samples to be skipped at each downsampling time in order to maintain the synchronization is computed from Eq.10. A backward shift is used to induce a delay in the antenna clock, while a forward shift gives an anticipation with respect to the rest signal.

just multiplying it by the sampling frequency. The correction is performed with the shifted downsampling depicted in Fig.1, including simultaneously both the spin-down and the Doppler effects. δt has to be computed only a few times over the observation time (each downsampling time), and the applied correction is valid for all the sources in the $\hat{n}$ direction, with the same value of the fractional spin-down rate ε, independently of the emission frequency. Typical downsampled frequencies range from the case of the targeted searches, where the emission frequency is known (a thousandths of Hz band and less is enough, corresponding to hundreds of s downsampling time), to semitargeted case, where a several Hz band (fraction of s downsampling time) is required. Slightly different methods to compute sampling shifts (Eq.10) in the two cases are used in these two scenarios [3].

4. Validation of the pipeline

Several tests of the resampling method have been performed by simulating signals with different frequency ν_0 and spin-down rates $\dot{\nu}_0$, considering several source sky directions, and signal sampling frequencies ν_s. In order to test the robustness of the method to trajectory changes, different parameters for the antenna orbit have been used, including the real Virgo antenna trajectory during its first scientific (VSR1, May-Oct 2007). The amplitude modulation due to the variation of the response of the antenna, induced by the different orientations during the Earth rotation [4], has not been included in the analysis. As shown in [5], this can be done at the end of the pipeline, on the final FFT, after the Doppler and spin-down correction. Both the phase locking accuracy and the peak recovering have been tested by simulating only the signal affected by Doppler and spin down, with no noise injected in the data stream. Spin-down rates ranging from $\dot{\nu}_0 = -10^{-12}$ Hz/s up to non-physical values ($\dot{\nu}_0 = -10^{-4}$ Hz/s) have been taken into account. The different tests will be discussed in a better detail in a forthcoming paper [3].

In Fig.2, the comparison of the phase of the modulated signal, coming from a fixed direction of the sky, with the rest one is performed before and after the correction.

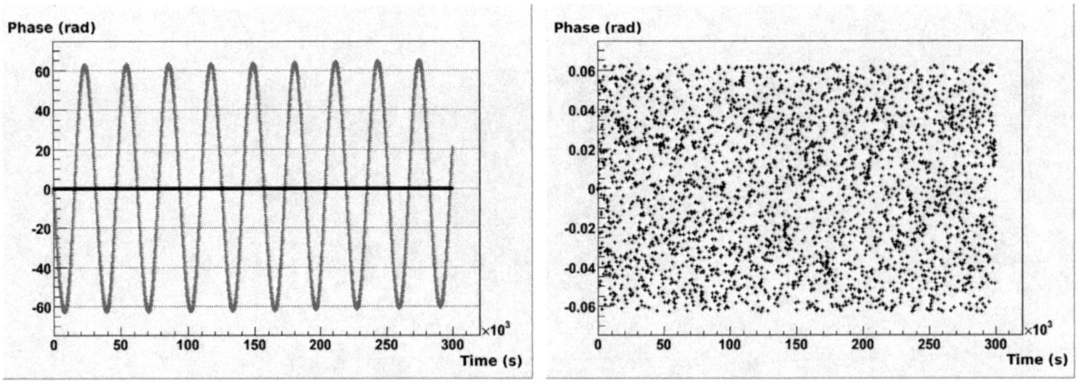

Figure 2. Phase locking accuracy test using a circular orbit around the Sun (with a radius of $1.5 \cdot 10^8$ m and a uniform orbital velocity of $\beta = 10^{-4}$) with $\nu_0 = 20$ Hz, $\dot{\nu}_0 = -10^{-11}$ Hz/s and $\nu_s = 1000$ Hz. The plot on the left shows, as a function of time, the phase difference between the rest signal and the modulated one (red curve), and the phase difference (close to zero) between rest and corrected signals (black curve). The large Doppler modulation of the phase induced by the circular orbit and the very small drift due to spin-down can be appreciated in the red curve. The residual dephasing after the correction, visible in the $y - axis$ zoom of the black curve (right plot), never exceeds the expected value, $\pm\pi\nu_0/\nu_s = \pm 0.063$ rad.

In all the tested cases the signal phase has been correctly recovered, with a residual dephasing with respect to the rest signal within the expected limit ($\pm\pi\nu_0/\nu_s$). One of the end-to-end tests performed on the semi-targeted pipeline application is illustrated in Fig.3 and its caption. We tested the accuracy in the peak reconstruction applying the technique to signals with different frequency, sampling frequency and spin-down rates. As shown in 4, the amplitude reconstruction accuracy results to depend, as expected, only on the ratio between the signal frequency and the sampling frequency (and not independently on the signal frequency, and on the spin-down rate). One can also observe that, once the signal frequency is one fifth of the sampling frequency, an acceptable level of reconstruction (amplitude losses well below 10 per cent) is achieved. This means that to investigate signals below 200 Hz (as usual in semi-targeted searches), 1 kHz sampling rate is enough.

The full pipeline for semi-targeted searches works as follows. The high-frequency signal (4 kHz or 20 kHz) is pass-banded around a given central frequency with a large band (a few Hz). The chosen bandwidth depends only on the maximum dimension of the FFT one wants to perform, and not by our method that is valid at the same time for all the emission frequencies (small enough with respect to the sampling frequency - see Fig.4). Indeed, after the correction taking place on the pass-banded stream, still at high frequency, one can heterodyne the signal, and thus makes a down-sampling of the signal at a frequency around the double of the bandwidth (again at a few Hz). The correction is valid for all the frequencies of the small band and thus, even considering a long integration time (months), one can perform the FFT on the full set of data, since, after the downsampling, we have a reasonable number of points. For instance, considering 120 days of integration time, and investigating a band whose width is 2 Hz (downsampling frequency of 5 Hz), after the correction (for a given spin-down rate), one can make an FFT on around 50 millions of points (small enough for effective FFT algorithms).

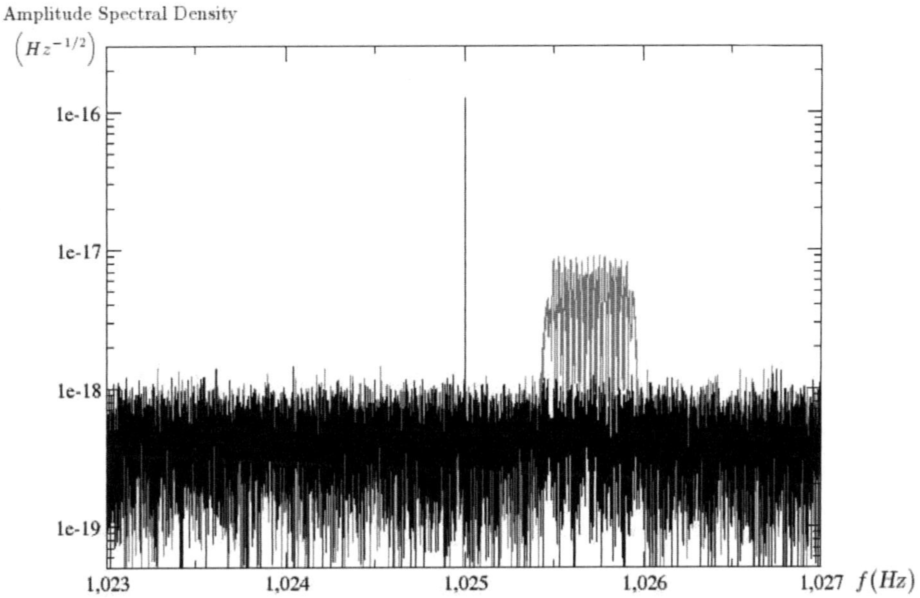

Figure 3. Linear spectral amplitude of the signal before (*gray*) and after (*black*) correction. A 30 days-long data stream, sampled at 4 kHz and made by random noise, with an amplitude similar to the Virgo one (in the concerning band), is filtered by a Chebyshev pass-band ($\nu_{stop} = 2.5$ Hz and central frequency $\nu_c = 19.0$ Hz). A gravitational wave signal, coming from the center of the galaxy, as it would be detected by the moving Virgo antenna (with the same orbital motion of VSR1), is injected in the data stream, before the pass-band filtering at $\nu = 20.025$ Hz, with $\dot{\nu}_0/\nu \simeq -5 \cdot 10^{-13}\ s^{-1}$, and amplitude $h_0 = 10^{-19}$. The smearing of the spectral energy induced by Doppler effect is visible in the gray curve, such as the recovering of the signal peak at 1.025 Hz in the black one. The losses with respect to the original signal are, after the reconstruction, below 1%.

In order to avoid ultra-long FFT, it is more effective to perform a broad-band semi-targeted search, splitting the analysis on different smaller bands, and repeating the method several times for different "spectral slices". The computational costs for each steps of the pipeline have been measured on a standard computer. The cost of the resampling has a negligible impact on the entire pipeline (10^{-5} of the length of the period investigated), with respect to the data reading (I/O) ($4 \cdot 10^{-4}$ of the period, for data input at 4 kHz). Details on the optimization of the pipeline for searches on many directions of the sky will be published in [3].

Conclusions

The proposed resampling method to compensate for Doppler and spin-down effect has been successfully tested. Good performance in terms of accuracy in spectral peak recovering, absence of artifacts in the noise, and fast execution time has been achieved. Since the Doppler and the spin-down correction, for a given fractional spin-down rate and direction, is valid simultaneously for all the investigated frequencies, the method results to be very effective in

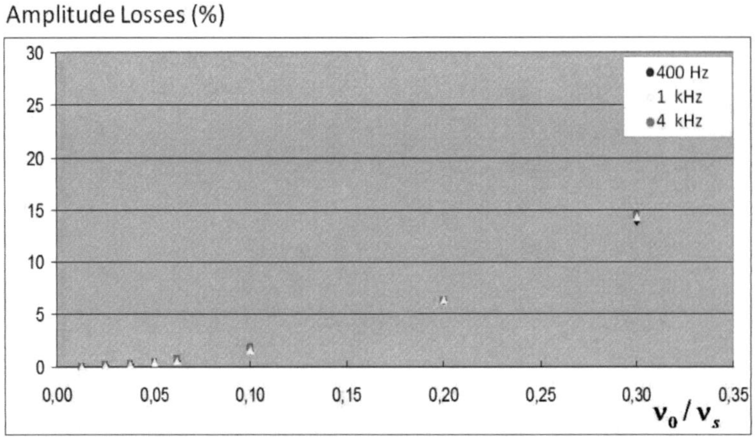

Figure 4. Peak amplitude losses as a function of the source-sampling frequency ratio (ν_0/ν_s). The plot is repeated for three different sampling frequencies (400 Hz, 1 kHz, 4 kHz). The losses, as expected, depend only on the frequency ratio, and not directly on the source frequency. The estimated errors are smaller than the plot resolution.

semitargeted searches, where the emission frequency is unknown. An analysis based on this method, concerning the data of the two recent Virgo scientific runs is in progress by using the GRID environment.

Acknowledgments

The authors acknowledge the support of the Rome 1 Virgo group, in particular S.Frasca, C.Palomba, F.Antonucci and P.Astone, for their smart suggestions. Thanks are also due the Ligo Scientific Community-Virgo joint group aimed to Continuous Wave data analysis, for the discussion of the method and the many suggestions received.

Journal of Physics: Conference Series **243** (2010) 012009 doi:10.1088/1742-6596/243/1/012009

References

[1] P.Jaranowski, A.Krolak, B.Schutz, *Phys.Rev. D* 58, 063001 (1998).
[2] See Pulsar catalogue of *Australian Telescope National Facility* at web site: *http://www.atnf.csiro.au/research/pulsar/psrcat/*
[3] S.Braccini, G.Cella, I.Ferrante, D.Passuello, O.Torre, *"Resampling technique to correct for Doppler effect in gravitational wave search"*, submitted to Physical Review D (2010).
[4] M.Maggiore, "Gravitational Waves, Volume 1: Theory and Experiments", Oxford University Press, USA (2007).
[5] S.Frasca, on behalf of the *Virgo* collaboration, "*The search for Vela Pulsar in Virgo VSR1 data*", to be published in *Proceedings of the 8th Amaldi conference on gravitational waves* (June 22-26, 2009), Columbia University, New York.

AUTHOR INDEX

AUTHOR INDEX

Institute of Physics
Dirac House, Temple Back
Bristol BS1 6BE UK